MASALA LAB
The Science of Indian Cooking

VIKING

MASALA LAB

Krish Ashok is not a chef but cooks daily. He is not a scientist, but he can explain science with easy-to-understand clarity. He trained to be an electronic engineer but is now a software engineer. He learnt to cook from the women in his family, who can make perfectly fluffy idli without lecturing people on lactobacilli and pH levels. He likes the scientific method not because it offers him the ability to bully people with knowledge, but because it confidently lets him say, 'I don't know, let me test it for myself.' When he is not cooking, he's usually playing subversive music on the violin or cello.

He lives in Chennai with his wife, who sagely prevents him from buying more gadgets for the kitchen, and his son, who is a biryani connoisseur.

You can follow him on Instagram, @_masalalab, and on YouTube, @krishashok, at your own risk.

MASALA LAB
The Science of Indian Cooking

Krish Ashok

PENGUIN
VIKING

An imprint of Penguin Random House

VIKING

USA | Canada | UK | Ireland | Australia
New Zealand | India | South Africa | China | Singapore

Viking is part of the Penguin Random House group of companies
whose addresses can be found at global.penguinrandomhouse.com

Published by Penguin Random House India Pvt. Ltd
4th Floor, Capital Tower 1, MG Road,
Gurugram 122 002, Haryana, India

Penguin
Random House
India

First published in Penguin Books by Penguin Random House India 2020
This edition published in Viking by Penguin Random House India 2023

Copyright © Krish Ashok 2020

Book design and illustrations by Meghna Menon

ISBN 9780670098569

Typeset in Sabon by Manipal Technologies Limited, Manipal
Printed at Thomson Press India Ltd, New Delhi

www.penguin.co.in

To Sonu,
who makes amazing rasam without any science

Before I left for my first trip abroad, I asked my late maternal grandmother, who was a fantastic cook, to tell me the recipe for adai, a crispy multi-lentil pancake that is very easy to make but hard to get right. I probed her about ratios, texture, timing and sequence. Clearly, she was not used to being asked these questions. Cooking for her came from aromas, the tactile memory of her fingers, and the visual and auditory cues. Once I was done with my interrogation, she asked me to show her what I had written. She said, 'You missed one ingredient. Write it down.'

I looked at her, pen in hand, waiting.

'Patience. That's the ingredient you are missing. If you give anything enough time, it will turn out delicious. You can approximate all the other ingredients.'

Contents

Introduction

Cooking, people will tell you, is an art. Indian cooking, in particular, is supposed to be an art wrapped in oriental mystique, soaked in exotic history and deep-fried in tradition and culture. Western food is supposed to be scientific and bland, while Indian cooking, we are told, is all about tradition and flavour. Some people innately have a knack for it, and many don't. The Tamil expression *kai manam*, which literally means 'hand flavour', is used as a compliment for those who have somehow had this arcane knowledge handed down to them. The metaphor also hints at where exactly that knowledge is stored (hint: not the brain) and, thus, is hard to transfer to another person.

The Covid-19 pandemic transformed life as we knew it. The fact that cooking is an essential life skill stares us more intensely in the eye than at any time so far. But learning to cook Indian food, it turns out, is a byzantine maze with conflicting instructions and pseudoscience. The convenience of being able to Swiggy some amazing butter chicken from a dhaba 5 km away, and the fact that more and more young people are living by themselves in cities, which are not their home towns, means that even if they want to cook, they neither have the time nor the daily access to someone who can mentor them in the way your grandmother learnt to cook, from an older member in her family. And because we have never bothered to build a standard, documented model of underlying cooking methods and the science behind those techniques, a metamodel, if you will, Indian cooking continues to wrongly be considered all art and no craft.

This is a pity, given that this part of the world has contributed traditional culinary methods the scientific West has embraced as new-age food science in recent years. The curcumin in turmeric is now a superfood, as is the drumstick, which is sold as moringa powder in Brooklyn for an arm and a leg. Fermentation and sprouting of legumes go back thousands of years in India, while pickling as a technique to extend the shelf life of food in a hot and unforgiving climate has been around forever. There is no dearth of hey-look-our-ancient-tradition-is-now-science chest-thumping on social media, but what is missing is any serious attempt at actually documenting these culinary practices as part of a practical engineering playbook, outside of the cultural, historical and spiritual contexts.

By treating our culinary tradition as something sacred, artistic and borderline spiritual, we are doing it a grave disservice. Let me take music as a metaphor here. Indian classical music, one of the most sophisticated artistic traditions in the world, has, I would argue, suffered from the lack of documentation and archiving. In fact, the insistence on purely oral traditions of transmission of knowledge have ended up making the art a very elitist affair not accessible to the wider population. Western music, in contrast, has a simple, visual system of notation that is able to accurately capture every nuance. Because of this, we are able to perform a Bach concerto in exactly the same way as he intended in the eighteenth century. As an amateur musician myself, my teachers would often tell me that

Indian classical music cannot be described and documented because its nuances are beyond the ability of language to describe it with fidelity. With due respect, I think that's bullshit. What we are doing with food is rather similar. By not using the tools and language of modern science and engineering to continuously analyse and document different Indian culinary traditions, and instead just writing down recipes, we are doing the food equivalent of lip-syncing to a pre-recorded track.

Food, even sun-blushed, Himalayan pink salt-tossed palak chaat, drizzled with organic, hand-blended coriander pesto, is ultimately chemistry. It's not always simple chemistry, I'll give you that, but then again, the physics of how the global positioning system (GPS) works on your smartphone involves Einstein's general theory of relativity, something that most Silicon Valley software nerds who build these apps don't understand entirely. I want to make a strong case for understanding basic food science, minus the chemistry equations and dry academic white papers on the correlation of temperature and water absorption levels in chickpeas (chickpeas soaked in warm water absorb it faster, a trick you can use in case you forgot to soak them overnight). Cooking is essentially chemical engineering in a home laboratory, known as a kitchen, with an optional lab coat, known as an apron. Unfortunately, pseudoscience, amplified by social media, has given the word 'chemical' a negative connotation. People regularly say, 'I don't want to eat anything that has chemicals in it.' In that case, I'd advise them to fast indefinitely. It's a cognitive fallacy, which assumes that somehow the glutamate salt in monosodium glutamate (MSG) is a chemical, while the same glutamates inside the fleshy part of a tomato are natural. At a molecular level, they are the same thing! But more on that in Chapter 5.

Understanding basic food science will make you a significantly better cook and also unlock your ability to experiment boldly without being tethered to recipes. Food science will also make for richer and more fascinating conversations with your grandmother when she teaches you how to make the perfect prawn theeyal. Knowing how pressure caramelization works, or rice to water absorption ratios, or the ability of glutamate molecules to add savouriness will help you imbibe her cooking methods and adapt it to many other dishes.

I can see why this sounds daunting, considering that we live in a world buried under the collective weight of flowery adjectives used in food literature in general. Tomatoes are either sun-blushed or French-kissed, the lowly black urad dal takes on a Silk Route ambience in the form of Bukhara or Samarkand, and spices are always exotic and recipes handed down over the ages. Indian food writing continues to romanticize desi cooking with the same orientalist tropes that look down on technology and promote an utterly fraudulent notion of 'natural and organic', continuing to perpetrate the silly idea that Indian cooking is not scientific and that those who are good cooks just know by '*shudh* desi ghee' instinct what to add, and how and when.

But food is ultimately just chemicals. And the process of turning a chickpea pod into a mouth-watering chana masala is engineering. It's about time that we take an engineering approach to cooking, in addition to the tradition, history and art approach. This is not to undervalue the history and art of food. Sure, I'd love to learn the fact that Emperor Akbar was particularly fond of murg jalfrezi while playing chess with Birbal, but that won't help me make a better murg jalfrezi. Akbar, it turns out, was mostly vegetarian, but you get my point. Learning to cook by reading recipes is like trying to learn chemistry by only reading equations, or the biographies of chemists. Imagine walking into a chemistry lab and finding an instruction textbook that says, 'Bring to a gentle simmer the exotic melange of oxalic acid and the riotous colours of potassium permanganate.' You would probably respond with, 'Can we go a little easy on the adjectives and stick to explaining how and why things work?'

This book is an attempt to de-exoticize Indian cooking and view it through the lens of food science and engineering. There is art in cooking, no doubt, but there is a lot of craft too, and craft is, as Michael Ruhlman puts it in his fantastic book *Ratio: The Simple Codes behind the Craft of Everyday Cooking*, founded on fundamentals. In this specific case, it would be the fundamentals of food chemistry. But this is not to invalidate your mother's traditional method of making sambar. In fact, it is aimed at explaining why, for instance, she dry-roasts fenugreek (methi) for a shorter period of time (releasing volatile flavour molecules) than the urad dal (Maillard reaction), or why she squeezes a bit of lime into the pot of simmering dal (acids make things taste more interesting). And since you are not your mother, it makes better sense for you to learn the science behind why she does what she does and apply it right away, rather than learning it yourself through years of cooking each day. As Harold McGee put it in *Keys to Good Cooking: A Guide to Making the Best of Foods and Recipes*, traditionally schooled cooks may not know chemistry, but they know cooking, and in the kitchen, it's the knowledge that counts. But if you do not have that knowledge, you have two choices: spend decades experimenting with food and figuring it out yourself, or do what humankind does better than any other species on the planet, translate collective wisdom into documented, tacit, practical knowledge on the basis of science. This book is an attempt to do just that for Indian cooking.

This book is also not about 'authentic' Indian cooking. Only completely fraudulent people swear by authenticity when it comes to food. What is an authentic sambar, really? My maternal grandmother, who was a great cook, grew up in a village near Tiruchirappalli (Trichy) in Tamil Nadu in the 1930s, a time when carrots, beans, cauliflowers and the likes were termed English vegetables, a term not uncommon in rural Tamil Nadu even today. They were not available to those living in a small village given the logistical complexities of transporting them from where they grew, the colder climes of Ooty. If you have sambar today, say in a restaurant in Chennai, it's quite likely that it will feature carrots. My grandmother grew up making sambar without carrots, but she started using them once she moved to Chennai, where they were available all through the year. More interestingly, if you go back a few hundred years, Indian cooking did not even include chillies and tomatoes—both of which came from the New World and were introduced to India by the Portuguese. So, if anyone gives you grief about 'authenticity' in food, please move them to the part of your brain labelled 'Recycle Bin' (and click on the 'Empty Recycle Bin' button for good measure).

Food science is about understanding what happens when different ingredients—spices, proteins, carbohydrates and fats—interact with each other at various temperatures, proportions and pressures. Also, 'flavour' is a multidimensional experience involving taste, aroma, mouthfeel (texture), sight and sound, and how understanding this can arm you with a set of techniques that will make your guests swoon over your chana masala.

So, how is this book structured? For that, let's consider the aforementioned chana masala. It begins with a chickpea, an annual (the plant germinates and produces the seed in a single growing season, after which it dies and has to be replanted) legume that is notoriously hard to cook because of the amount of dietary fibre the seed packs in the form of cellulose and hemicellulose. Soaking it for six to eight hours causes the water to seep into the seed via osmosis and leads it to expand in size. You could try and cook it without soaking, but unless you know some food science, you will likely fail. I'll explain why.

At this point, you have two options. You can cook it in an open vessel for several hours and waste cooking fuel, or you can pressure cook. You can then look up a recipe online that asks you to pressure cook the chana for eight whistles. At this point, you must rip your Internet connection cable and dump your computer into the nearest dustbin, because everything you thought you knew about pressure cooking is wrong, and literally every recipe online is indulging in wholesale misinformation. Chapter 1 will explain the scientific way to go about pressure cooking, and you will discover a whole new universe of things you can do in the kitchen with the pressure cooker, beyond cooking lentils and rice.

Coming back to our dish, you will then add the soaked chana, whole spices like black cardamom and cloves, and a pinch of baking soda to the pressure cooker. Sodium bicarbonate is, despite its bad reputation due to overuse by low-cost restaurants, one of the most magical ingredients in the kitchen when used appropriately. Soda reacts with the pectin (a hemicellulose) in the skin of the chickpea, which breaks it down faster than by just applying heat and pressure. Using a pinch of soda is, therefore, both energy efficient and results in a perfectly soft final product. Did you know that baking soda can also be used to accelerate the Maillard reaction, which produces the delicious browning effect on food? Chapter 5 will teach you about the many other tricks this simple ingredient can pull off to produce the most astonishing and delicious outcomes.

But that's not it. The food science enthusiast (and your grandmother and mother) will not stop at just the soda. It turns out that sodium bicarbonate is mildly basic (as opposed to acidic) and bases tend to taste bitter. And when soda reacts with an acid (like the hydrochloric acid in our stomachs), your high-school chemistry lessons should remind you that it produces carbon dioxide, a gas that many of us regularly expel in the form of a burp or a fart. An interesting aside: We also breathe out carbon dioxide every other second, and that's how our bodies lose weight, and by breathing it in, plants gain weight. But back to our sodium bicarbonate now. When there is too much unused soda in the food, you get a feeling of being bloated, which is the main reason why we tend to consider soda as bad. That's where food science comes in. Chapter 4 explains the role of acids in Indian cooking. Lime juice, yoghurt, tamarind juice and vinegar are all examples of acids. Fun fact: Even sulphuric acid, with its gory reputation thanks to Bollywood, is used in the food industry to make cheese.

Basically, anything sour is an acid. In the case of the pressure-cooked chana, however, we use another mild acid for a very practical reason—a teabag. Tea is mildly acidic, but crucially, it is inactive till it is heated in water, unlike lime juice which immediately reacts with anything it fancies. Adding a teabag to the chana has two advantages. It neutralizes all the unused baking soda, so that your stomach does not have to do it for you (burrp!) and lends a lovely dark brown colour to the chana. Of course, the teabag also keeps the bitter leaves out of the chana.

So, while the chana is being pressure-cooked, we start with the gravy. Every recipe book will give you a different list of ingredients, and every garam masala/chana masala spice mix will have unique ingredients. Here's the thing: None of that matters. Just see what your kitchen has and use it. Modern Indian cooking is more about technique than about the quality of ingredients. Sure, premium ingredients help, but technique matters way more than the quality. This is because urban Indians rarely have access to high-quality fresh produce, unless they are quite rich. Contrast this with Italian cuisine, where most dishes use a tiny number of really high-quality ingredients, and the difference between Italian food made with ordinary ingredients and premium ingredients is like that between day and night.

Spices are absolutely central to Indian food, and yet, most people waste their hard-earned money by failing to understand the basic chemistry of flavour molecules and their volatility. Do not waste good money on buying a giant box of chana masala spice mix, which you will use only once in a few months. It will turn into flavourless sand as it oxidizes on your kitchen shelf, despite your best attempts at keeping those boxes airtight. Chapter 2 will teach you how to make any masala when you need it, absolutely fresh, and just the right quantity. Once you've made your spice mix, heat some ghee, add onions and a pinch of salt, and slow cook them till they brown. Not translucent, but brown. The difference in flavour between translucent onions and browned onions is massive. Chapter 3 will unlock the secrets of the Maillard reaction, which is the difference between blandness and incredible depth of flavour in any ingredient.

Once the onions are suitably browned, you should typically add chopped chillies, ginger and garlic paste, and tomatoes too, and if you are smart, a tiny bit of concentrated tomato paste. If you don't have tomato paste, you could just add one of those ketchup sachets saved up from home deliveries. Tomato ketchup is a fantastically underrated flavour enhancer. Chapter 4 will help you use acids in your cooking.

Back to the recipe. At this point, you add the chana masala spice mix. Flavour molecules dissolve best in hot oil or alcohol, not water. So, before you add water to your gravy, always add a splash of the cheapest brandy or vodka you can find and, trust me, the heat will burn off most of the alcohol by the time the dish is done, but not before it extracts a ton of more flavour from all the spices. Chapter 5 will explain how to use alcohol, an ingredient used regularly in Italian and French cooking, in Indian cooking.

You might also wonder what combination of ingredients should go into your gravy. Should you use potatoes? Can you make this without garlic or onion? How do you thicken the gravy? Rather than being tethered to a recipe, this book aims to liberate you from their tyranny by taking an algorithmic, software-engineering approach. Chapter 7 will give you an algorithm for generating gravy, rice, bread, chutney and salad recipes. You want to make your chana masala Punjabi style? Use ghee, cumin (jeera), onions, tomatoes, amchoor (a spice powder made from dried unripe green mangoes), and ginger and garlic. Malabar style? Use shallots, garlic, curry leaves, coconut milk. Bengali style? Use mustard oil, nigella seeds, mustard, fenugreek, fennel and carom seeds (ajwain). If someone tells you it isn't authentic enough, ask them to take a hike. We aren't looking for *authentic*. We are looking for *delicious* and *adventurous*.

Once the gravy is done cooking and you open the pressure cooker, discard the teabag and whole spices (people who leave cooked husks of whole spices in their dishes must be sentenced to life imprisonment. There is no flavour left in whole spices after being pressure cooked for 15 minutes, trust me) and add the chana to the gravy. Bring it to a simmer, at which point you could also add anardana (dry pomegranate seeds powder, with an acidic profile that adds sourness) or vinegar-pickled ginger, where the vinegar serves two purposes—muting the sharp heat of ginger, while adding a fresh and vibrant sourness. Another flavour enhancer is sugar (preferably brown). A teaspoon of sugar will elevate the taste of any dish.

If you are wondering how I qualify to write a book on this topic, all I can say is that I am a software engineer by profession. If you are uncharitable, you can treat this as yet another attempt by a 'tech bro' to wade into a subject without credentials. If you are charitable, you can consider the fact that good engineering is about optimizing, simplifying and standardizing processes. I cook daily and have zero dietary restrictions (other than eating endangered species). I'm not a restaurateur or chef, but I regularly chat up chefs in small restaurants to learn how they achieve consistency of taste every single time, and also how they rustle up a dal makhani in under 10 minutes. My immediate goal is to explain food science in simple, non-technical terms that everyone can intuitively understand. My long-term, perhaps overly ambitious, goal is to build a community of Indian food science enthusiasts who will build on this book and encourage everyone to experiment with newer methods of cooking Indian food, and also accurately document the stunning variety of cooking methods in this part of the world, beyond mere recipes. No two recipes for a dish tend to be the same in India, and if every single one of these claims to be authentic, then authenticity as a concept is meaningless in this context.

In summary, this book is an attempt to introduce an engineering mindset into Indian cooking, with the ultimate aim of making the reader a better cook and turning the kitchen into a joyful and creative laboratory for culinary experiments. It also aims to equip you with modular nuggets of scientific knowledge, which will help you adapt and invent new dishes with greater facility. At the same time, it is important to say this. A vast majority of home cooking in India is done by women, with no help and little choice in the matter. Urging someone who just wants to feed the family in the shortest possible time, while balancing family and career, to carefully consider whether the citral in lemongrass pairs well with the piperine in pepper before using them together is an exercise in dilettantism, like trying to upsell a laborious chore as a hobby. That I get to treat my kitchen as a laboratory is a privilege not many have. My hope is that this book has enough simple science and engineering lessons to help cut down cooking time, while improving flavour and predictability.

Chapter 1

Cooking is ultimately the application of heat to physically and chemically transform ingredients into food. Understanding the basic physics of heat and the chemistry of water are essential to becoming a better cook. How you apply heat to grains, legumes, vegetables, meat, eggs and fat, and understanding what happens to food physically and chemically at different temperatures will give you a greater degree of mastery and control over cooking. Natural, experienced cooks tend to know these things intuitively, but most of that knowledge is tacit and not documented in simple, scientific terms. For e.g., almost everyone in India gets pressure cooking wrong. The first chapter will clarify the basic science behind pressure cooking and teach you how to get it right every time, while saving on energy costs. Pressure cooking can produce an astonishing range of flavours when done right. If you thought it was just about rice and dal, be prepared to be blown away.

Chapter 2

Did you know that you can only taste things that are soluble in water? That almost all flavour molecules are not water-soluble? It also turns out that most powdered spices lose their flavour in a couple of weeks, so that frayed box of garam masala on your kitchen counter likely adds little or no flavour to your dal tadka. This chapter will reveal to you the secret to forever-fresh and flavourful spices: a low-cost coffee grinder. This chapter will offer a simple, intuitive way to understand where flavour comes from, how to extract the right amount from your spices and how to combine flavours in the best possible way. It will lay out for you regional and common spice mix templates, which will let you prepare any masala from any part of India just before you make a dish.

Chapter 3

This chapter will focus on cooking's most famous reaction— the Maillard reaction—and describe it in the context of two of the most commonly used flavouring ingredients in Indian cooking: onion and garlic. Most dishes in India start with these two. How you cook them, and how you can extract even more flavour out of them, is the aim of this chapter. Beyond onions and garlic, this chapter will also give you tips on browning anything to boost its flavour, including plain old, boring cabbage. And finally, we will look at the most extreme use of the Maillard reaction in the kitchen—deep-frying—and how to get it right every single time.

Chapter 4

Here we will explore the role of pH levels and how understanding it can turn bland-tasting dishes into vibrant disco parties for your taste buds. When you add yoghurt, tamarind, vinegar or lime juice to a dish, you are effectively adding an acid, which is what produces a tangy or sour taste. But there are more culinary acids than just these. Knowing how to use them will unlock a universe of new flavours.

Chapter 5

While the first half of the book focuses on things that most people know the practical applications of, this chapter will focus on ingredients that have unfortunately been smeared by pseudoscience for decades but are precocious magic wands in the kitchen. MSG and sodium bicarbonate have a wide range of uses in the kitchen, if you are willing to consider the fact that there is zero scientific evidence of them being harmful, provided they are used in small amounts. This chapter will also introduce you to the idea of using alcohol when cooking, a practice quite common in the West.

Chapter 6

This chapter will introduce the idea of modernist Indian cooking—the use of new cooking techniques and equipment that will not just save you time but also improve the flavour of your dishes significantly. We will explore interesting uses of the microwave oven, beyond heating water and reheating food alone; look at how electronic pressure cookers (like the Instant Pot) work; modernist ingredients like sodium citrate, xanthan gum and liquid smoke; and also attempt to find an answer to the age-old question of how to cook chicken breast without it becoming rubbery and dry, i.e., sous vide cooking.

Chapter 7

This chapter will attempt to convince you that recipes are a limiting way to approach cooking. Instead, it will introduce you to the concept of a metamodel for the common styles of Indian dishes. We will describe repeatable and scalable algorithms for dishes that use regional ingredients, and a common set of modular cooking techniques to help you make any gravy, rice dish, bread, salad or chutney from any part of India. It will also teach you kitchen time-optimization techniques like preparing base gravies that you can freeze and later being able to whip up a dish in less than 10–15 minutes. This is what restaurants do, and there is no reason you should not do this at home. It is also a great way to ensure that fresh produce doesn't go bad sitting in your refrigerator.

Chapter 8

Contemporary India has had an uneasy relationship with invaders from the north-west, but one particular invasion seems to have avoided controversy for the most part—biryani. This delectable juxtaposition of the subcontinent's staple grain (rice) with meat and spices has conquered the nation, all the way from the delicate Awadhi biryani to the intensely spicy, yet minimalist, Ambur biryani 2000 km south of Lucknow. This chapter will recap every single lesson from the previous chapters to help you make biryani at home—be it keeping the meat (or vegetables) juicy, the rice moist yet perfectly cooked and separated, and the spices perfectly balanced without being overwhelming. Those who make good biryani get to enjoy that most unique pleasure of watching a guest take a mouthful and swoon over the song of the endorphins cavorting about in their brains, as their facial expressions scream, 'Thank you for this delicious meal.' Yes, good biryani (and all those endorphins) is ultimately chemistry.

And oh! While other books on cooking feature gorgeously plated food photographed with high-end SLR cameras, this book features illustrations by me that look like they were drawn by a middle schooler.

The illustrations in the book are by Meghna—an illustrator and graphic designer by day; cat butler, connoisseur of instant noodles, video game nerd and creepy-crawly enthusiast by night. She draws a lot of fan art and hopes to one day write and illustrate her own graphic novel! You can find her on Instagram and Tumblr, @soulstuffjunkie.

1 Zero-Pressure Cooking

Time is an illusion. Lunchtime, doubly so.

—Douglas Adams

What Is Cooking?

A kitchen in an Indian home is, notwithstanding the sterile and spotless ones featured on cooking shows on TV and YouTube channels, a chaotic command centre overseeing a war. This war is a daily strategic mission to turn ingredients wrapped in plastic bags, secured with extra rubber bands and stored in airtight containers, into freshly cooked food, all the while keeping at bay a horde of microorganisms steroid-pumped by the warm and tropical climate, seeking to consume these ingredients. There are multiple fronts in this war. Despite the presence of a refrigerator in most homes, there is a distinctly patriarchal, and thus religiously sanctioned, preference for eating freshly cooked food, which is usually made by the women from scratch, using manual labour as much as possible. Hence, grains and legumes have to be pressure-cooked, vegetables peeled and chopped, and spice powders used before they turn into flavourless sand. The level of efficient and continuous partial attention required to do this day in and day out, without burning food or losing blood to dull knives, tends to overwhelm the average newcomer into the kitchen. Most beginner cooks don't get consistent-enough results from their efforts in a reasonably short amount of time for them to fall in love with cooking, and that, I think, is a pity. Art is and should be hard to master, but if the craft is hard to get right, then the documentation is probably inadequate

Cooking is ultimately chemical engineering. If factories can produce complex consumer products through the cunning use of automation, supply chains and skill-specialization, home cooking can be transformed into an engineering discipline, one with standardized methods and a finite set of easy-to-remember general principles that work for all kinds of dishes.

Cooking is the transfer of energy, usually in the form of heat, to your food:

1. To change the physical structure of the carbohydrates, proteins and fats in it. For instance, what happens when you boil a potato?

2. To modify the water content of ingredients. What happens when you deep-fry a cutlet?

3. To speed up chemical reactions that make food more flavourful and easy to digest. What happens when you pan-fry shrimp till it is slightly golden brown?

4. And it's typically one-way. You can't un-fry fried fish.

Food science is the biology, chemistry and physics relevant to turning other forms of life into food that nourishes us, while also tasting delicious. The processed food industry is the largest employer of food scientists whose job is to enable the industrial manufacture of safe, nutritious (okay, not always) and appetizing food. It involves a discipline at the intersection of biology, chemical engineering and industrial automation. This book has a humble and narrow scope—to bring just the optimal amount of scientific rigour and engineering ideas to the home kitchen, and more critically, providing you with a set of principles that will significantly increase the chances of you cooking something delicious every time you walk into the kitchen. This book will not make you a chef. It will make you a better home cook. And even if it doesn't do that, it will, at the very least, make you curious about how delicious food is made. Hopefully, that will make you fall in love with the craft of cooking

Basic Physics of Cooking

Let's start with the single most crucial process that turns ingredients into food. Heat. More specifically, the transfer of heat from an energy source, like a stove, an induction stovetop, or an oven, to your food. Heat, physicists will tell you, is energy, but that doesn't really clarify things, so let's step back and understand what energy is. That, it turns out, is hard to define with first principles, but I've found Nathan Myhrvold's definition in *Modernist Cuisine* a rather useful abstraction—energy is the ability to make things happen. It is easier to understand energy in terms of what it does rather than what it is. There are many kinds of energies: the one in the nucleus of an atom (which makes a nuclear bomb possible) and the energy in the bonds between atoms, but heat is best understood as a measure of the energy from the continuous, random movement and collision of atoms and molecules in a substance. In food-relevant terms, the potato in your hand is made up of carbohydrates, proteins and water molecules, all of which are moving around all the time, even in the couchest of couch potatoes. When you drop it into boiling water, where the water molecules are moving around even faster, because it has more energy, the simple principle is that heat will move from a place where there is more to a place where there is less of it. In the kitchen, stoves and ovens are sources of energy that can physically and chemically modify food, while refrigerators remove energy from your food and stall most chemical reactions. If you are wondering where all that energy goes, just feel the back of your fridge.

And that brings us to the concept of temperature. It is simply a quantification of how much heat there is in a substance or system. But here's a tricky thing to consider. If you touch a pan that is at 70°C, you

0°C   Water freezes at sea level

24°C   Coconut oil melts

29°C   Chennai declares winter

50°C   Proteins in meat begin to denature

62°C   Eggs begin to set

68°C   Collagen in meat begins to denature

70°C   Vegetable starches begin to break down

154°C   Maillard reaction becomes noticeable

180°C   Sugar begins to caramelize visibly

480°C   Temperature in a tandoor oven

will burn your finger badly. If you touch a pot of water at 70°C, it will mildly scald you, but if you put your hand inside an oven at 70°C, the air will feel only mildly hot. So, just understanding temperature isn't enough because different substances at the same temperature feel different to us and, thus, have entirely different effects on food. We need to understand a few more concepts to get the hang of this, which brings us to density.

Density is a measure of how much substance there is in a given space. Intuitively, metals are denser than water and water is denser than air. Solids tend to be denser than liquids or gases. So, there is more solid in a given amount of space than there is liquid, which means there are more molecules in a solid than in a liquid in the same space. More molecules moving around means that there is more heat energy, which is why metal at 70°C feels hotter than water at 70°C. In fact, boiling water (at 100°C) has more heat than hot air at significantly higher temperatures, say, in an oven. At this point, it should dawn on you that this is precisely why cooking something dry on metal pans is the fastest way to cook (or burn) food, while boiling something in water takes more time, and baking in an oven takes even longer.

Let's understand a few more concepts around this idea. Specific heat capacity is the amount of energy required to raise the temperature of a substance. Water has high specific heat capacity. Steel, on the other hand, has very low specific heat capacity. So, even though there is more 'substance' in steel, which is solid, it takes less time to heat it up than water, a phenomenon we are all frustratingly familiar with when waiting for milk, which is mostly water, to boil. This is what we really mean when we say that metals are good conductors of heat.

Air pressure is the measure of the amount of force experienced by anything as a result of the weight of air above it. This is important to cooking because of how water behaves at low or high air pressures. The boiling point of water is the temperature at which there is enough energy to break the bonds between water molecules, which keep it in liquid state, and liberate them into the air as water vapour (steam). Intuitively, the water molecules have to deal with the pressure of the air above them to free themselves. Imagine trying to walk through a crowd of people. The larger the crowd, the harder it is to walk through. So, the higher the air pressure, more is the energy required to liberate the water molecules. This is why at high air pressures, the boiling point of water is more than 100°C, which means you need more heat to boil water. Conversely, at lower air pressures, the boiling point is less than 100°C. This is particularly common at high altitudes where there is literally less air. Air becomes thinner the higher you go, which is why planes need stored oxygen and cannot merely open some windows to let air in.

Now that we have understood some of the basic physics of heat, let's talk about the different mechanisms through which heat can be transferred from one substance to another. Conduction is the transfer of heat from a solid to any other substance via surface contact. Conduction is how the oil you add to a metal pan heats up as a result of energy transferred from the metal to the oil. If you go really deep, the molecules inside the metal are moving around at a furious pace because they have been heated up by the stove, and metals heat up really fast because they have low specific heat capacities. These molecules collide with the molecules in the oil and transfer some of their energy to them via these collisions. That's how the oil heats up.

Convection is the transfer of heat from a liquid or gas to your food. Because liquids and gases have molecules that are farther apart, the transfer of heat via conduction, where molecules close to one another happily collide and heat things up, is not as practical, so the transfer of heat in water happens via bulk movement of molecules from the hotter parts of the liquid to the cooler parts. These convection 'currents' then transfer heat to the things they come in contact with. An oven and a pot of boiling water operate on the same principle, while using different convection mediums.

The third mechanism for transferring heat is radiation. Electromagnetic waves with high-enough energies can transfer heat directly to your food, provided they are close enough to the source. The broiler in your oven works this way. Incidentally, placing something in the hot sun also heats it up using the exact same principle, except that the source of this radiation is 150 million km away, while the bulb in your oven is a few centimetres away from the food. A hot light source generates infrared radiation that heats up and cooks the food under it. You can see this in operation in a place that makes shawarma. A glowing source of heat generates infrared electromagnetic waves, and that heats a rotating spit packed with meat. Technically speaking, all objects, including us, radiate some amount of energy as electromagnetic waves, but most of that energy isn't powerful enough to cook food (thankfully). But if the object in question is reactor number 4 at Chernobyl, the results are not pretty.

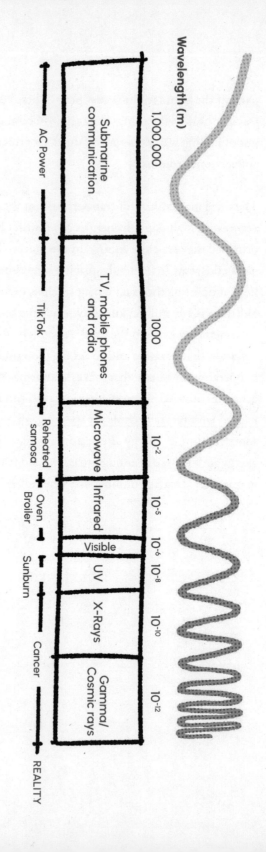

Technically speaking, microwaves are also a form of electromagnetic radiation. Why then, you might wonder, do they belong to their own category of heating methods? This is because microwaves by themselves don't have enough energy to heat any and all substances. In fact, microwaves have less energy than infrared waves that heat up slices of bread in a toaster. What they are good at is specifically causing water molecules to align themselves to the magnetic field they create. So, engineers figured out that if they keep changing the direction of the microwaves, the water molecules will keep flipping to align to the direction of the magnetic field. And since flipping is movement, and movement is energy, the water heats up. Microwaves work by heating up the water inside foods, which is why they don't work for food items that don't have enough moisture.

Basic Chemistry of Cooking

Now that you understand the physics of heat, it's time to understand what happens to the molecules in food when heat is applied. Here is the shortest possible chemistry lesson on how ingredients transform into food.

While there are a million different chemical reactions that happen when you cook food, the four major ones that are worth understanding are: starch gelatinization, protein denaturation, hydrolysis and the Maillard reaction. But before we get to those, let's do a simple lesson on what a chemical reaction is, and to do that we must start with molecules. Molecules are combinations of atoms that have formed bonds with each other until heat, enzymes, acids or bases do them apart. And what is a bond? The simplest explanation of a bond is that it is a transaction involving atoms greedy for electrons and atoms that are generous donors of electrons.

There are four kinds of bonds they form. The strongest is called an ionic bond, something you see in sodium chloride (salt). The second kind is the covalent bond, a slightly more flexible form of marriage, one that you see in water. The third kind is the hydrogen bond, again found in water, where the oxygen and hydrogen atoms form side relationships with other water molecules' hydrogen and oxygen atoms. The last kind of bond is the Van Der Waals bond, the weakest of the lot but crucial to cooking because fats use this bond to attain the viscous texture they have. This is critical to cooking, as we shall discover in subsequent chapters.

A chemical reaction is the formation and/or breaking of bonds in the molecules in your food. Many of these reactions comprise not single but hundreds of steps. The four that we are interested in are:

1. Starch Gelatinization: This is what happens when you cook starches in grains like rice, wheat, lentils, potatoes, yams, etc. In the presence of water and heat, long starch molecules break up and some parts of their molecules form hydrogen bonds with water. This is why rice or dal increase in size after being cooked in water. This happens between 55°C and 85°C, depending on the kind of starch.

2. Protein Denaturation: This is what happens when heat or acids are applied to proteins. The long and complex structure of proteins unfolds in such a case. You can see this reaction when you add vinegar to warm milk and watch it curdle into paneer. This is also the reaction that 'cooks' proteins into harder, less elastic meat. This happens around 60°C.

3. Hydrolysis: This is when proteins break down beyond just the unfolding that happens in the denaturation phase. This is the reaction that makes meat tender when connective tissues break down into gelatin. This can also be accelerated

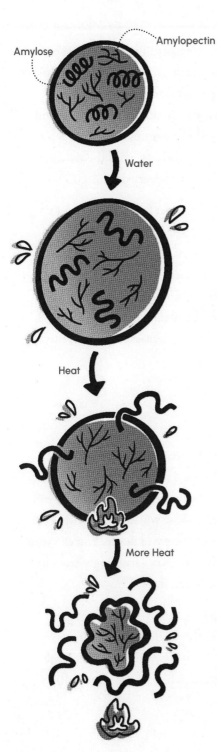

Amylose

Amylopectin

Raw starch granules are made up of two kinds of molecules: Amylose, which is linear and helical, and amylopectin, which is branched

Water

When water is added, the amylose molecules' structure is disrupted and water is absorbed, causing the starch granule to swell. This is what happens when you soak rice or legumes

Heat

When heat is applied, in addition to water, the amylose molecules now start escaping the granule

More Heat

As heat continues to be applied, all the amylose escapes the granules which now are mostly only made up of amylopectin. The amylose and water mixture becomes a soft gelatinous goo, a texture we are familiar with from cooked starch in rice or potato by the use of enzymes, which are basically proteins that can assist the hydrolysis reaction. Bromelain, found in pineapple juice, and papain, found in papayas, are used to 'tenderize' meat this way. Hydrolysis, which also happens to sugars and fats, is enabled by a strong acidic environment. This is what happens inside your stomach, by the way, with the two litres of hydrochloric acid that it produces every day.

4. Maillard Reaction: Arguably the most famous reaction in cooking, it causes browning of food and a host of delicious, aromatic by-products. This is what is happening when your onions brown or your chicken sears in a pan. This reaction happens between 110°C and 170°C. A close cousin of the Maillard reaction is enzymatic browning, which is not very desirable—this is what turns your potatoes and apples brown when they are left exposed to air.

Cooking Techniques

Now that you have armed yourself with the basic thermodynamics of cooking, and the four most common chemical reactions when you cook, it's time to get practical and understand the different ways in which you can apply heat to food. What method you choose will depend on the kind of ingredients you are cooking and the outcome you desire. You can turn a potato into a gravy to be had with puris, or into a satisfyingly crunchy chip based on how you choose to cut and cook it.

Broadly, there are two ways of cooking food: moist and dry. Moist cooking methods involve the use of water (in its liquid form) or water vapour. And since water boils at 100°C (unless you are in Shimla), this means that all moist cooking methods have to be executed under 100°C.

Once water turns into vapour, it tends to escape and shirk its cooking responsibilities, unless you are using a steamer or pressure cooker.

Moist cooking methods include:

1. Blanching: Brief exposure to boiling water, generally used for spinach and other delicate ingredients that don't survive sustained cooking.

2. Poaching: Cooking in water that is just short of bubbling (typically at 85°C).

3. Simmering: Cooking in water that is just bubbling but has not reached a full boil (90–95°C).

4. Boiling: Cooking in water that is close to turning into vapour. Typically used for hard, tough-to-cook ingredients that can take the violence of boiling water.

5. Steaming: Cooking using water vapour. Since vapour is less dense than water, steaming takes more time than boiling but leaches fewer nutrients into the water. Momos are typically steamed because dropping the delicate parcels into boiling water would cause them to disintegrate.

6. Braising: Cooking ingredients in just enough liquid, so that it becomes the sauce in the dish.

7. Stewing: Cooking with a liquid at low heat, used for tougher cuts of meat.

8. Pressure-cooking: Remember how the boiling point of water increases if air pressure goes up? A pressure cooker exploits this by creating an airtight enclosure where water boils along with your food. As the air pressure of the vapour keeps increasing, it prevents further vapour generation by

raising the boiling point. So, a pressure cooker can cook food at almost 121°C, which is why it cooks faster than water in a regular vessel.

Some things to remember about moist cooking methods:

1. You can't get any kind of browning done with moist cooking methods, and brown, as far as food is concerned, is the colour of magic. This book has an entire chapter dedicated to the Maillard reaction, the one that coaxes the most amazing flavour out of food.

2. Vegetables and meats respond to moist cooking methods differently. Vegetables and grains almost always get soft when cooked in water, and meats almost always get hard and dry if cooked the same way, unless you constantly keep the temperature below 70°C.

Dry cooking methods involve temperatures well above 100°C, when food rapidly dehydrates (loses water) and acquires delicious flavours and a brown colour as a result of the Maillard reaction. Dry cooking methods include:

1. Sautéing: The use of a hot metal surface, and a little fat, to cook food rapidly.

2. Roasting: The use of low heat applied over a long period of time to cook food slowly. Potatoes, for example, do better with roasting than sautéing. When done over hot coals, or on wood on top of a grill, it's called barbecuing.

3. Baking: The use of hot air in an oven to cook food.

4. Broiling: The use of infrared waves in an oven, or on a grill with hot coals underneath, to cook food.

5. Frying: The use of fat heated to around 170°C–190°C to rapidly dehydrate the surface of food, thus making it brittle yet non-porous. This allows the exterior to brown as a result of the Maillard reaction, while the insides cook more gently without structural collapse.

6. Smoking: The use of cold or hot smoke from burning specific kinds of wood to impart a rich flavour.

In addition to differentiating between moist and dry cooking methods, it is also important to understand the difference between slow and fast cooking methods. In slow cooking, small amounts of heat are applied for short periods of time. Deep-frying or baking in a tandoor are fast cooking methods, while cooking mutton for a biryani at low heat for 45 minutes is slow cooking.

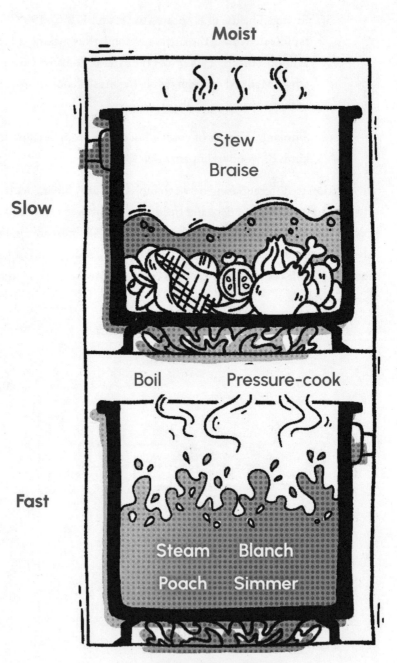

Moist

Stew
Braise

Slow

Boil   Pressure-cook

Fast

Steam   Blanch
Poach   Simmer

Less than 100°C

Dry

Roast

Slow

BBQ

Sauté   Bake   Fry

Broil

Fast

More than 100°C

Not all ingredients behave the same way when heat is applied. Understanding these differences can make you a better-informed cook. Here are a few examples:

1. Heat adds flavour to meat and fish until a point, after which they start to become dry and chewy. Cooking meat to be moist and tender, while ensuring it is perfectly cooked, takes precision.

2. Heat mutes the flavour of onion and garlic. The raw ingredient is too overpowering in itself, and the aim of cooking is to reduce, not amplify, its flavour. Given how ubiquitous these two ingredients are in Indian cooking, we have an entire chapter dedicated to them.

3. Heat significantly improves the flavour of cabbage, way more than it does for other vegetables.

4. Heat improves the flavour of tomato. In fact, the longer and slower you cook, the more amazing are the flavours that you can extract from a tomato. It is not uncommon in Italy to see pasta sauces being cooked for an entire day!

5. Heat adds bitterness to green leafy vegetables and, thus, must be applied very carefully.

As you read the rest of this book, you will discover more of these basic principles that will help you fine-tune your finesse in the kitchen

Materials

What your cooking vessels are made of makes a significant difference to how heat is applied to your food. The choice of material and its thickness will affect the time taken to cook, evenness of cooking and retention of heat. Aluminium, for instance, heats up really quickly but does not stay hot for long once you reduce the heat on your stove. This is good for applying high amounts of heat for short periods and then reducing it to prevent scorching. Aluminium will respond quicker to the act of reducing the heat. Stainless steel, on the other hand, takes time to heat up but stays hot for longer. This means that if you get it really hot and then want to reduce its temperature, hard luck. The steel will hold on to the heat for quite some time, risking your food getting overcooked.

The choice of material will also affect whether or not you can cook some ingredients. For instance, aluminium and cast iron will react with any acid in your food, like tomatoes or tamarind, and produce off-putting metallic flavours. Stainless steel, however, will not react with your food at all.

Non-stick cookware can be problematic too, especially if low-cost variants are used. The Teflon coating is somewhat sensitive to high heat and can disintegrate, getting into your food with repeated use. For example, what if you accidentally overheat a non-stick vessel on an induction hob, and believe me, this is easier to do than you think. Induction stoves do not generate heat on their surface, and thus you can't feel the heat like you do on a gas flame. The noxious fumes from the breakdown of the non-stick material are poisonous enough to kill a small bird. A ceramic non-stick skillet is a safer option, as that accidental overheating will not kill little birds. But you can damage a ceramic non-stick surface quite easily with aggressive washing or scrubbing.

Material	Properties
Aluminium	Good conductor, does not retain heat well, reacts with acids, cheap.
Anodized Aluminium	Advantages of aluminium,, plus does not react with acids. Costlier than aluminium.
Stainless Steel	Not a good conductor, retains heat, does not react with acids. Price depends on quality of steel.
Non-stick	Non-stick surface sensitive to high heat. When overheated, fumes can kill small birds.'
Ceramic Non-stick	Non-stick surface does not last long, but overheating will not cause deaths of small birds.
Seasoned Cast Iron	Retains heat insanely well. Reacts with acids. Weighs a ton.
Tri-ply	Combination of stainless steel and aluminium's advantages, Costs an arm and a leg.
Copper	Amazing conductor of heat, reacts with acids and costs your entire extended family's arms and legs.

	Dry	Gravy
Sour	Khatte Aloo/ Bhindi	Fish curry, Sambar, Gongura, Vindaloo, Kadhi/Yoghurt
Not Sour	Potato fry, fish fry, baked food, foil-wrap, tadka pan	Shukto, Khorma, Boiling Milk

My recommendation is to keep a ceramic non-stick skillet just for cooking eggs, because eggs will stick to literally any surface other than non-stick or really well-seasoned cast iron. If you get the pan temperature precisely right, and the hen that laid your egg did yoga and Pilates, you can manage to cook eggs without too much sticking. But in my experience, it's not worth it.

Here's a bare minimum list of vessels for the typical urban Indian kitchen:

1. Kadai: A hemispherical vessel not suitable for induction stoves, but ideal for deep-frying and stir-frying because the lack of any corners ensures that no food gets stuck in a hard-to-reach edge. The shape of the vessel naturally makes it hold less oil at the bottom, where it's narrower.

2. Flat-bottomed frying pan for making dry dishes.

3. Saucepan with tall vertical edges for gravies.

4. Pressure cooker or pressure pan.

This apart, I would also strongly recommend some other vessels:

1. Stock Pot: A large-sized, thick-bottomed vessel with a lid, which can be used to make stocks, as well as biryani, besides being used as a water bath for sous vide cooking (see Chapter 6).

2. Steamer (or Steaming Stand): To use vapour to cook delicate food like momos and idlis, or cook vegetables without losing out on nutrients.

As for kitchen tools, here is a short list of absolutely essential items for the science-minded home cook:

1. Instant-Read Thermometer: The key to transforming ingredients into perfectly cooked food is precise application of heat, and precision requires measurement. An instant-read thermometer will tell you if the oil is hot enough to make crisp, fluffy puris, or if it will result in soggy, greasy ones. Also, meat is notoriously sensitive to temperature, and this will help you achieve consistent results in terms of tenderness and doneness.

2. Weighing Scale: Since you are not familiar with the innate touch and feel of dough elasticity the way your grandmother, or mother, is, you don't know how much water to add to how much atta to make the perfectly soft chapattis consistently. So, get into the habit of weighing ingredients and making notes for consistency. Also, get out of the habit of measuring by volume (cups and tablespoons). Weights (in grams) are more consistent and accurate.

3. Tongs: If you do not possess the precise flipping skills needed to flame-grill a chapatti, a pair of tongs will help you. It will also help you pick up things being deep-fried with precision.

4. Silicone Spatulas: Instead of noisily scraping a metal ladle against a metal vessel, trying to pour out every last bit of that delicious dal, use a silicone spatula. There is also a commonly held notion that silicone is not heat resistant. It can, in fact, be used up to 200°C safely. And if you are cooking at that high a temperature, you might have bigger problems to worry about than melted silicone.

5. Mandoline Slicer: This will help you slice vegetables precisely, which is particularly useful when making chips at home, and also slicing through onions before they assault your eyes.

6. Citrus Juicer: This helps filter out citrus seeds that are insanely bitter. Those who clumsily use their hands to squeeze limes, and then fail and filter out all the seeds that have fallen into their dishes must be taken to the woods and shot.

7. Spice Grinder: This will come in handy because store-bought spice mixes turn into flavourless sand soon after you open them. There is an entire chapter dedicated to this (see Chapter 2).

8. Microplane Grater: This is the most efficient way to make fresh ginger and garlic paste. Store-bought paste tastes nasty because of the sodium citrate and other preservatives in it.

Citrus juicer

Mandoline slicer

Instant-read
thermometer

Digital weighing
scale

Spice (actually
coffee) grinder

Heat Sources

A gas burner stove connected to a liquefied petroleum gas (LPG) cylinder is still the most common heat source in an urban Indian kitchen. An entire generation of cooks has, in fact, grown up getting used to three heat settings: sim, medium and high. Medium serves as the most commonly used setting to cook food, while high is used occasionally for deep-frying or pressure-cooking. Sim is for taking a short break from the kitchen without causing the food to burn.

But now that induction hobs are becoming popular, everyone except experienced, natural cooks are failing to realize that the heat transfer mechanism in an induction stove is entirely different from that in a gas stove. If you have cooked on a gas flame all your life, switching to an induction stove needs a slight reset of a lot of muscle memory. In a regular stove, the gas burns, heating up the vessel through direct contact. If you remember the physics from a few paragraphs earlier, the flame causes molecules in the pan to vibrate. Since solids are dense, vibrating molecules pass on their vibrations to their neighbours and, in a material that is a good conductor, the neighbours pick up vibrations fast and without complaints. Once the pan is heated, the food in it starts to get hot because its molecules are in contact with the disco vibration party in the pan, and disco is pretty infectious.

But in an induction stove, an alternating current flowing through a copper coil creates an oscillating magnetic field in its vicinity. When a cooking vessel made of a magnetic material (cast iron or steel) is placed on the stove, the oscillating magnetic field induces an electric current, which flows around in a loop, into the pan. In the universe we live in, electrical fields induce magnetic fields and vice versa, and a current flowing through any material runs into electrical resistance, which produces heat. In simple terms, the pan literally heats itself because currents induced into it by the wireless, oscillating magnetic field face resistance from its own material. This is why when you turn on an induction stove, the top, which is typically made of a non-magnetic, poor heat-conducting ceramic material, does not feel hot at all! And because no conduction of heat is involved, the vessel heats up really fast. In fact, water will boil 30 to 40 per cent faster on an induction stove compared to a traditional stove.

So, when you use an induction stove, the amount of heat induced in your cooking vessel is dependent on the amount of electrical power supplied to the copper coil inside. This is usually measured in watts. Here's a simple heuristic: 300 W is medium–low, 500–600 W is medium–high (the most commonly used setting) and anything above that may burn your food pretty quickly, so use it only when bringing water to a boil, or bringing a pressure cooker to peak pressure.

The third most common heat source in the kitchen is the microwave oven, which works by heating water inside the food. Since most food contains water, you can pretty much heat most edible things, of course, with some limitations. It's great for reheating food, heating water, melting butter, and if you are really lazy and creative, making entire meals for one. Because it is such an under-utilized device in the Indian kitchen, and also a victim of all manners of it-is-dangerous-radiation pseudoscientific fearmongering, we will discuss the microwave oven in great detail in Chapter 6.

The last, and the least common, heat source in the typical Indian urban kitchen is the convection oven, whose mini-me version is the oven-toaster-grill (OTG) that is slightly more popular than full-sized ovens. OTGs and convection ovens use heated air (convection) and infrared radiation (broiling) to cook food and tend to be used only by people who bake bread or cakes. But they are also an excellent way to brown food without using too much oil. An air fryer is very similar to a convection oven, except it uses a tiny compartment that allows it to brown and crisp food even quicker.

A convection oven is the most practical way to cook at really high temperatures (well above 100°C and up to 230°C) in a controlled manner. You can place something on a pan heated to 200°C, but that will scorch the food because, as we learnt just a while back, metal at 200°C has way more energy than hot air at the same temperature. Thus, the amount of energy transferred to your food by hot air is much less, so you can gently cook something up to a high temperature without burning it. A tandoor, which is also a convection oven, goes up to 450°C, which is blazing hot and can cook things in a very short period of time. This allows it to retain moisture better in food than home ovens, which tend to dry things out over long bake times. Water, if you remember, boils at 100°C.

The Magic of Water

Most things you cook probably have a lot of water in them. Most fruits and vegetables are more than 80 per cent water by weight. It is sometimes unintuitive to think that a carrot has roughly the same proportion of water as milk (around 88 per cent) and a cucumber contains more water (95 per cent) than the hard, mineral-heavy tap water you are likely to get in an Indian city. In fact, most food is mostly water with a small amount of proteins, fats, carbohydrates, minerals and vitamins. So, if you don't understand water, you can't understand cooking.

Consider a water molecule. It is, despite its deceptive simplicity, an extraordinary substance. On the face of it, it is two hydrogen atoms bonded to an oxygen atom (H_2O), but that's a bit like describing a Raja Ravi Varma painting in terms of the shades of paint he uses. For starters, it is extraordinary that it is liquid at temperatures human beings tend to be comfortable in. Most substances similar to water in chemical structure and simplicity, and not made of metals, are gases! Carbon dioxide is a good example, as is hydrogen sulphide (H_2S) that gives chaat masala its characteristic smell. In fact, sulphur is in the same family of elements as oxygen, just heavier, and yet, at room temperature, H_2O is liquid and H_2S is a gas!

Without getting into advanced chemistry, let's try and understand this visually. Oxygen, an atom that is rather greedy for electrons, forms a V-shaped bond with hydrogen, which gives up its electrons rather easily, making the water molecule 'polarized' (meaning that the oxygen side of things is negatively charged). In contrast, the hydrogen side is positively charged. Since opposite charges attract, the oxygen of one water molecule is also attracted to the hydrogen of nearby molecules, in blatant violation of the commandment about not coveting thy neighbour's atoms. So, what this does is that it makes water molecules stick to one another, which gives water this interesting property of high cohesion, an effect you can see in dew drops on leaves.

In liquid water, H_2O molecules constantly form and break hydrogen bonds like the characters in Archie comics. This allows water to be in liquid form at room temperature

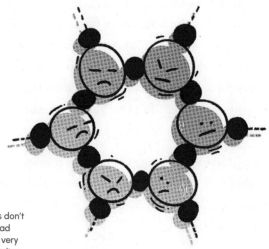

But as temperature falls, water molecules don't get to cavort around with abandon. Instead they end up forming hydrogen bonds in a very specific orientation with strict military discipline. This is why ice floats on water.

Another astonishing property of water is that its solid form is less dense than its liquid form! Normally, solids tend to be denser than liquids, given that is pretty much the dictionary definition of 'solid'.

Since the hydrogen and oxygen atoms in every water molecule form bonds with nearby water molecules' H and O atoms (hydrogen bonds, if you recall), it turns out that in liquid water the molecules tend to have, on an average, more energy than what is required to break one of these hydrogen bonds. So, liquid water is this chaos of hydrogen bonds forming and breaking all the time, like the high-school dating scene in an American teenage drama. But as we lower the temperature, the water molecules have less energy to keep moving around. Around 0°C, the hydrogen bonds lock into place in a very precise order and do not break. It is because of this strict arrangement that ice is less dense than water.

This property, among other things, makes water absolutely critical to life. We are mostly water, and most of what we eat is also mostly water.

When scientists look for life on other celestial bodies, the first thing they try to find is water. Life, even a fundamentally alien form of it, is unimaginable without water for, among several other reasons, the fact that its liquid form makes every scientist go 'no way should this be liquid'. We could write an entire book on just liquid water and its astonishing characteristics that make life, as we know it, possible, but let's stick to what makes it special in the kitchen.

Water is critical to the texture of food. Its presence makes vegetables crisp, in that they wilt when they lose water, and meat tender, in that meat becomes dry and hard without it, while a potato chip tastes best in its relative absence.

When you add salt to water—sodium chloride, which is also very strongly polarized as a molecule because chlorine is even more greedy for electrons than oxygen—the sodium and chloride ions turn that illicit three-way relationship between oxygen, hydrogen and a nearby molecule's hydrogen into a late Roman Empire-style orgy involving sodium, chlorine, oxygen and hydrogen ions.

This is, in the simplest sense, what 'dissolving' something in water means. When you add something like oil to water, you will see the water literally push the oil away like it coughed loudly without wearing a mask. This is because the oil molecules are not charged enough to be able to break the cohesive forces of liquid water.

Water also has a high specific heat capacity, meaning that it takes a fair bit of energy to heat it up in the first place, an annoyance everyone who has spilt milk on the stove understands rather well. Fun fact: This feature, rather important in the kitchen, is also critical to the planet. It's because we have so much water on the surface that a place like Mumbai has mostly predictable temperatures, while Bhopal can swing wildly not just across seasons but in a single day. Anything with more water cooks slower than anything with less water because of its specific heat capacity. This is why garlic, which has less water by proportion, must be added after the onions have cooked, or else it will likely burn.

Which brings us to cooking. Water is indispensable to cooking for a simple reason. The ingredients we cook are all organic substances, which means they don't take well to high heat applied very rapidly, and our cooking vessels are almost always made of materials that conduct heat very well. When you add ingredients to a cooking vessel that is being heated on top of a stove, the part of the food that is in contact with the vessel is experiencing a tremendous transfer of heat compared to the part that isn't. This is because air is a terrible conductor. So, what happens here is that your food ends up burning before it has a chance to cook evenly. This is where the magic of water comes in. Just add a little water to your vessel and your ingredients will experience even heat transfer from all directions.

Remember the boiling point of water—the temperature at which the individual molecules of water are vibrating and moving around so fast that those oxygen and third-party hydrogen relationships are not tenable. Things have, literally, got too hot to handle. At this point, the individual molecules break away from one another and water turns into vapour.

This happens at 100°C when you are at sea level. If the water boils away, your food will burn, so one has no choice but to keep the cooking temperature under 100°C. This is why cooking rice or lentils in an open vessel seems to take forever. Here is where air pressure comes into the picture. If you try and cook rice in an open vessel in Shimla you will be somewhat frustrated because water boils at 92°C there. The air pressure in Shimla, which is 7500 feet above sea level, is lower than it is in Chennai. You can intuitively visualize this. Air pressure is literally the amount of force that the air around you is exerting on you. If you are liquid water that is being heated up, the more the pressure, the harder it is for the individual molecules to break apart and float away as vapour. Denis Papin realized that if he could find a way to artificially increase the air pressure inside a vessel, he could raise the boiling point of water and, thus, reduce cooking time. Thus was the pressure cooker born

Pressure-Cooking

The foundational principle of a pressure cooker is to cook food in water at high pressure. You add water to a vessel and tightly seal it so that no air escapes, and then heat it from below. The water starts to boil, and some of it becomes vapour, but now this vapour is trapped and cannot escape. This increases air pressure inside the vessel because there is now more gas trapped in it, and this pressure prevents more water from turning into vapour. And, of course, engineers figured out that we also need a safety valve to make sure that the pressure inside does not rise to a point high enough for the cooker to turn into a bomb. This safety valve is designed as an opening at the top of the sealed vessel, on top of which a calibrated weight rests. The mass of this 'weight' is calibrated in such a way that if the pressure increases beyond a certain limit, it will overcome the gravitational force exerted on the weight and cause it to move up and open the valve, letting go off some of the steam. This reduces the pressure inside the vessel and the weight falls back down. The general idea is to try and keep the pressure inside approximately 1 bar above air pressure. At this increased air pressure, water boils at 121°C. So, a pressure cooker is, in its simplest sense, a device that lets you cook using liquid water at 121°C (unlike an open vessel that allows you to do this at 100°C).

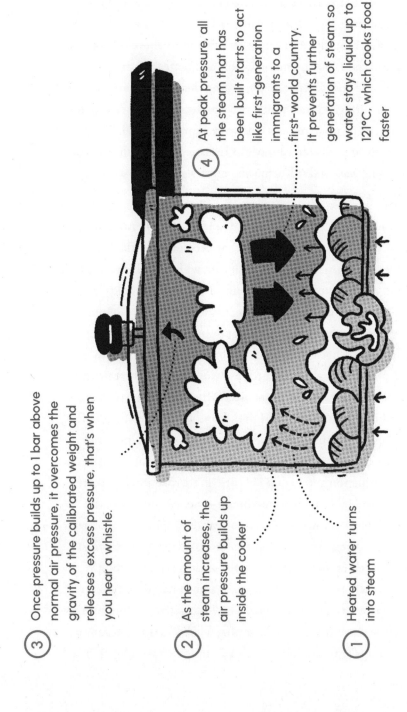

③ Once pressure builds up to 1 bar above normal air pressure, it overcomes the gravity of the calibrated weight and releases excess pressure, that's when you hear a whistle.

② As the amount of steam increases, the air pressure builds up inside the cooker

① Heated water turns into steam

④ At peak pressure, all the steam that has been built starts to act like first-generation immigrants to a first-world country. It prevents further generation of steam so water stays liquid up to 121°C, which cooks food faster

A temperature of 121°C is hot enough to cut cooking time by 30–40 per cent in most situations. But there is a rather interesting problem that I've noticed more recently. A fair number of people believe that pressure cookers made in the 1980s are more reliable and better constructed than the ones being sold today. Apparently, during the golden era of pressure cooker manufacturing, cookers would make perfectly cooked rice in three whistles, whereas the modern ones tend to be temperamental. This is not true. What is true is that a fair number of Indians are using the pressure cooker the wrong way, and still achieving mostly good results, so we never bother to correct our understanding.

The entire confusion stems from this widespread misunderstanding of how pressure cookers work. A significant percentage of the Indian population measures pressure-cooking time in whistles. This worked reliably for an entire generation because most urban households had more or less the same standard two-burner stove in the pre-economic liberalization era. So, the amount of heat put out by the stove at high, medium and sim settings was more or less in the same range across houses. The pressure would build up, you would hear the first whistle and some excess pressure would be released. After a while, it would build up to maximum pressure again and the second whistle would go off, and so on. This just fortuitously worked well enough for typical rice and dal cooking. But nowadays, we have a dizzying diversity in stoves, with electric, induction and high BTU (British thermal unit) burners that can rustle up Hakka noodles restaurant-style. At this point, the whistle method breaks down because it only works if the amount of heat you are applying is constant. This problem is particularly acute when using induction stoves. At its highest energy setting, typically 2000 W, an induction stove will bring water to a boil in almost half the amount of time that it will take a regular gas flame. This means that pressure cookers on an induction stove at the 2000 W setting will build up to maximum pressure, blow a whistle and rebuild to maximum pressure again in a significantly lesser amount of time than on a gas stove. If you cook rice in a pressure cooker at 2000 W on an induction stove, and measure in whistles, you are guaranteed to get undercooked rice. Or for that matter, dal or potato. You can, however, opt for two ways: reduce the heat setting on the induction stove to 500–600 W, so that the whistle count works, or learn the actual science of how pressure-cooking works.

The scientific, common-sense way to measure pressure-cooking to achieve consistency in results is the actual elapsed time at maximum pressure. Once a pressure cooker comes to full pressure and releases some steam, that's when the clock starts. Here is a quick guide to determine the pressure-cooking time for various ingredients:

While pressure-cooking is a tremendous time-saver, and in the case of a few green vegetables, moderately better at retaining nutrition and colour, I don't recommend throwing every single ingredient into a pressure cooker and turning it into a homogeneous mush in the single-minded pursuit of cooking dishes in one shot. While you can make a decent sambar and a half-decent biryani or pulao in a pressure cooker, you cannot make a great sambar or pulao because great cooking requires flavour-layering and textural variations, which are impossible to achieve by assaulting every ingredient with uniform 121°C heat at high pressure. But if you are short on time, there is no better method.

That said, try and avoid pressure-cooking for meats and seafood. Rather counter-intuitively, meats tend to dry out even with moist cooking methods. The key to great-tasting meat is to ensure that its internal temperature never goes above 70°C. In a literal pressure-cooker atmosphere, that is simply not possible, so the chances that you will end up with dry, overcooked pieces is very high. However, it is not uncommon to use pressure-cooking with tougher cuts of red meat, such as beef or mutton, to save time. But if you are looking to get the best flavour, low and slow is the way to go.

Vegetables		Legumes	
Beans	3 mins	Moong	6 mins
Beetroot (cubed)	6 mins	Toor	8 mins
Broccoli Florets	2 mins	Chana dal	15 mins (soak 30 mins)
Cauliflower Florets	2 mins	Chickpea	20 mins (soak 8 hours)
Carrot chunks	4 mins	Black urad	20 mins (soak 8 hours)
Potato (small)	6 mins		
Potato (large)	10 mins	Rajma (small)	20 mins (soak 8 hours)
Colocasia	3 mins		
Tapioca chunks	8 mins		

Millets		Meats	
Foxtail Millet	Soak: 3-4 hours Cook: 10-12 mins	Chicken (bone in)	8 mins
Little Millet	Soak: 2-3 hours Cook: 8-10 mins	Beef, Pork, Mutton (softer cuts)	15 mins
Kodo Millet	Soak: 3-4 hours Cook: 10-12 mins	Beef, Pork, Mutton (harder cuts)	40 mins
Barnyard Millet	Soak: 2-3 hours Cook: 8-10 mins		
Proso Millet	Soak: 2-3 hours Cook: 8-10 mins		
Pear Millet	Soak: 4-6 hours Cook: 12-15 mins		
Finger Millet	Soak: 4-6 hours Cook: 10-12 mins		

Science of Rice

Rice feeds more people on the planet than any other food. This comes down largely to the fact that the vast majority of people in India and China eat rice every single day. So, after we have dealt with the thermodynamics of heat, the material science of vessels, the extraordinary magic of water and the efficiencies of pressure-cooking, it's only fair that we turn our eye to this fantastic grain that grows on a tallish grass and, with the right application of water, heat and flavouring, turns into the subcontinent's magnificent culinary creation, the biryani.

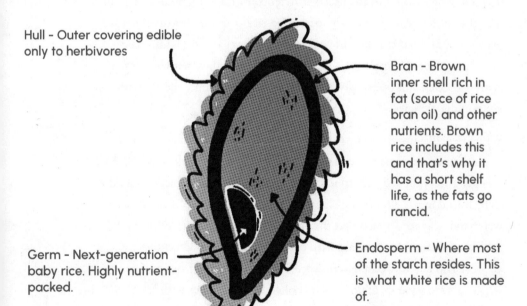

Hull - Outer covering edible only to herbivores

Bran - Brown inner shell rich in fat (source of rice bran oil) and other nutrients. Brown rice includes this and that's why it has a short shelf life, as the fats go rancid.

Germ - Next-generation baby rice. Highly nutrient-packed.

Endosperm - Where most of the starch resides. This is what white rice is made of.

Rice is designed to nourish the next generation of rice plants, much like an egg is designed to nourish the next generation of chickens. The germ is the next baby rice plant. The husk, which is the rough protective layer for the grain, is inedible unless you are a cow, and while the bran is highly nutritious and contains proteins and fat (the source of rice bran oil), it goes rancid quickly. This is why brown rice, which is rice with the bran and germ included, has a short shelf life. So, if you buy brown rice, don't buy large amounts. Refrigerate it if you do. When the bran is removed, we get white rice, which has a fantastic shelf life. That is what we mostly eat. Because white rice is just the starchy endosperm with no other nutrients, the process of parboiling the whole grain is used to enrich the nutrition of the endosperm. When the whole grain is partially boiled, it drives a good amount of the useful nutrients found in the bran and germ into the endosperm. This is why parboiled rice is almost as nutritious as brown rice and has the shelf life of regular white rice. This is also why most heavy rice-eating parts of India tend to have a diet that includes both parboiled and polished white rice, so that they don't end up with vitamin deficiencies.

White rice, like most starchy foods, has two kinds of starch molecules: amylose and amylopectin. Amylose is a smaller, linear molecule; amylopectin is larger and branched-out. In uncooked starch, amylose molecules tend to be found inside concentric chains of amylopectin molecules. It's when water and heat are applied that the amylose breaks through to form a gel-like substance we associate with cooked starch. The percentage of amylose and amylopectin determine if a certain variety of rice will be sticky or have separate grains when cooked. Rice varieties with less than 20 per cent amylose (80 per cent amylopectin) tend to become a little sticky after cooking, while varieties with more than 20 per cent amylose tend to have separate grains.

1. Basmati rice is aromatic and tends to contain between 20 and 25 per cent amylose, which makes it suitable for pulao and biryani, where you want separated grains post cooking.

2. Ponni rice from south India has 16 per cent amylose, which makes it slightly stickier than basmati.

3. Gobhindobhog rice from Bengal is aromatic and has about 18 per cent amylose.

4. Sona Masoori rice has 23 per cent amylose and is non-aromatic.

When rice is cooked in hot water, a process called gelatinization happens. This is where starch molecules, which are made up of long chains of sugar molecules, break down and form cosy relationships with water molecules to create two kinds of textures—a hard and waxy texture from the amylose, and a sticky gooey texture from the amylopectin. So, in addition to the variety of rice you use, your cooking methods also go a long way in determining if your final product will be nicely separated and fluffy, or sticky.

The general, fail-safe algorithm to cook rice perfectly is:

1. Wash away as much amylopectin as possible from the surface of the rice. This is the starch that becomes sticky when cooked. Wash rice till the water runs clear.

2. Then add water to the rice and bring it to a boil. When the rice's internal temperature hits 65°C, starches gelatinize. Don't worry, you don't have to poke a thermometer into a rice grain to accurately measure this. Just let the water visibly come to a boil, and the moment that happens you will know that all the starch in your rice has gelatinized.

3. As the water comes to a boil, place a lid on your cooking vessel and reduce the heat to the lowest setting possible. At this point, we are simply letting the gelatinized starches absorb the rest of the water in the vessel. This takes about 15 minutes.

4. Once all the water has visibly been absorbed, take your vessel off the stove and let it sit for 10 more minutes with the lid closed. At this point, a process called retrogradation happens, where each grain separates and creates its own identity, much like a teenager reading Ayn Rand. Once this is done, open the lid and fluff up the rice with a fork before serving.

Two questions arise here: How much water should you use? And isn't a pressure cooker a more convenient way to do this? Let's address the tricky water issue first. Everyone who tells you that you should use a 1:X ratio of rice to water is giving you partially correct information based on the wrong reasoning. Perfectly cooked rice is rice that has typically absorbed water in a 1:1 ratio by volume. Anything less and it will taste powdery and dry. And if you let it cook with more water, it will keep absorbing water and turn into congee. But if you add water in a 1:1 ratio, you will end up with undercooked rice because a good amount of that water will evaporate when it comes to a boil. So, it is necessary to add extra water to compensate for evaporation. That's the tricky part. Estimating how much extra water you need on top of the 1:1 ratio requires some experimentation with the vessel you use. This can also be a matter of personal preference. I, for instance, have grown up eating rice made with a 1:2 ratio, which is mushier than pulao but works perfectly with sambar or rasam.

Here is where your grandmother's till-the-first-knuckle-of-your-index-finger rule is good science. Even celebrity chefs on TV, who make it seem like the 1:2 ratio is linearly scalable, get this wrong. If you are using a narrow, vertical-walled vessel, and not a wide vessel where the evaporation rate is higher, the general rule is to add water till it is as high as one knuckle of your index finger above the rice level. For small amounts of rice, it just fortuitously turns out that a ratio between 1:1.5 and 1:2 works out to be close to the first knuckle rule followed for small, nuclear family-sized vessels. Thinking in ratios is also why a sizeable number of people mess up rice when cooking for a party. If you are cooking two cups of rice, four cups of water is way too much, but one knuckle above the rice level will always work because, that way, you are thinking in terms of compensating for evaporation, and the amount of evaporation does not depend on the amount of rice. It only depends on the amount of water above the rice level, which will be same regardless of the amount of rice you use.

The second question is: Why all this hassle? Isn't pressure-cooking rice more convenient? The answer is yes. If convenience is the primary concern, pressure-cooked rice is perfectly serviceable, but it will not be the tastiest and fluffiest of rice you can make. Only an open vessel and the strategic application of high and low heat will get you perfectly cooked rice. As someone who grew up eating pressure-cooked rice (and still does, for the most part), it's also worth considering that for a culinary tradition that makes astonishingly sophisticated and flavour-bomby side dishes, taking the time and effort to make perfectly cooked plain rice seems like an overkill. And I tend to agree. For most day-to-day purposes, pressure-cooked rice is perfectly fine. But on days when you feel like taking this grain from a tallish grass and turning it into the pinnacle of perfect texture, mouthfeel and taste, follow the method above.

Science of Lentils

A common misconception about lentils is that they are rich in proteins. In general, plants don't focus on making proteins the same way animals do. The proteins plants make are typically nutritionally incomplete for humans.

But, to be fair, lentils are packed with a lot of the good stuff, particularly harder-to-digest carbohydrates, which makes them a good source of plant-based protein in a balanced meal. And there are legumes richer in protein than toor dal. Fun fact: Two of the hard-to-digest carbohydrates in legumes like kidney beans (rajma)—raffinose and stachyose—cannot be digested by our digestive systems efficiently and, thus, become food for the bacteria in our guts. They metabolize these carbohydrates and produce gas, causing a rather familiar discomfort and occasional wind production. Turns out, eating fart-producing beans is not a bad idea at all because it encourages the growth of a diverse colony of healthy gut bacteria, who are, in general, excellent tenants.

Some lentils can be hard to cook and require a fair amount of time. Soaking reduces cooking time significantly. Though soaking does technically leach some flavour into the water, the difference is largely imperceptible because we tend to add a ton of extra flavour using spices.

Pressure-cooking also helps to shave off cooking times by almost 50 per cent. One of the hardest legumes to cook, the chickpea (chana), can be cooked to perfect softness if you add a pinch of baking soda to the pressure cooker. Baking soda (see Chapter 5) breaks down pectin, the hard substance that holds the plant's cell walls together, and accelerates the cooking of chickpeas (or any other legume for that matter). As always, our knowledgeable grandmothers will also throw in a teabag into the pressure cooker when making chana. They might tell you that it's meant to impart a lovely dark brown colour to the pale white chana, but the more useful, non-cosmetic purpose is to neutralize all the unused baking soda, which has a nasty, bitter and soapy aftertaste. Tea, as we will learn in Chapter 4, is an acid, while baking soda is basic. Acids and bases tend to react and neutralize each other.

Another minor annoyance when cooking dal is the foam it produces in the pressure cooker, which makes it hard to clean the lid afterwards. A teaspoon of oil added to the water in the pressure cooker will significantly reduce foaming when cooking legumes.

Urad dal in particular plays a big role in south Indian cooking. Lactobacteria on the surface of the dal and rice will, in the presence of water, cause fermentation, a behaviour exploited to make idlis, dosas and other lip-smacking items. Given that the weather is warm and humid all through the year in south India, fermentation is largely predictable and controllable.

Here is how you can make the perfect idli/dosa batter from scratch:

1. Take parboiled rice and decorticated (sounds cooler than saying de-skinned) urad dal in a 4:1 ratio (by volume), and soak them separately in water. Rice will need at least six hours of soaking, while urad dal will require just two hours. Don't oversoak the dal, or you will end up with a pasty texture in your idlis.

2. Grind the rice and dal separately with a pinch of fenugreek seeds, which have been soaked for an hour, and salt till you get a smooth texture. Don't over grind. Mix them together and let it ferment at room temperature. If you live in a colder place, heat up an oven, switch it off and then let it cool down to about 35°C before placing the batter inside.

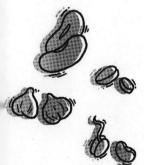

3. It will take between six and eight hours for the ideal amount of fermentation to happen before you can make idlis. The density of the batter will decrease as the bacteria eat the sugar in the rice and dal, and fart out carbon dioxide, which leavens the batter. Fermentation will also increase the amount of vitamin B in the batter and reduce its pH, thanks to the production of lactic acid that makes the batter mildly sour, a topic we shall explore in detail in Chapter 3. The amount of lactic acid will increase as the batter continues to ferment, which is why dosas are sourer than idlis and utthapams are the sourest of them all.

4. After six to eight hours, refrigerate your batter if you don't plan to use it right away.

5. You can make idli on the first two days, and as fermentation continues slowly in the refrigerator, you can make dosas on the third and fourth days, and utthapam after that.

6. Before using the batter, check its 'pourability'. It should feel like melted ice cream. If it feels thicker, add more water.

You can also use baking powder after soaking and grinding the batter to skip the fermentation process altogether, but you will then lose out on the complex depth of flavour that slow fermentation brings. Here is an important caveat: Since there are too many variables involved in fermentation—humidity, room temperature, quality of grains/lentils, general mood and inclination of the bacteria, etc.—here is a scientific way to arrive at the perfect method that works for you. Take small bits of rice and dal in different ratios. Try 3:1 and 4:1. And for each one, try a four-, six- and eight-hour fermentation time. Make idlis using each batch and see which combination's flavour and mouthfeel you like. Document that as your standard method. And when you store the batter in the fridge, put a sticker with the refrigeration date on it so that you know when it is best suited for idlis, dosas and utthapam.

Science of Wheat

Now that we have rice and dal out of the way, let's consider the other staple carbohydrate: wheat. The original grain that made large-scale human civilization possible, wheat (like rice, corn and sugarcane) is a grass, making the grass family of plants one of the most successful species on the planet. Whether we have domesticated these grasses, or they have deviously convinced human beings to stay addicted to carbohydrates and, thus, grow them on a massive scale, at the cost of other plants, is a question worth pondering over when you mix atta and water and let it sit for 30 minutes. If you aren't doing autolysis, which is what this step is called, you are skipping the single biggest science trick when it comes to making the perfect chapatti, or a paratha, naan or kulcha for that matter.

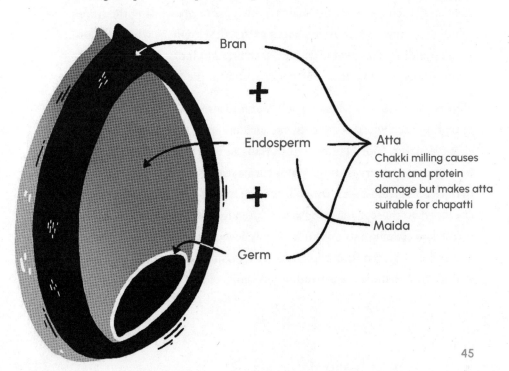

Bran

+

Endosperm

+

Germ

Atta
Chakki milling causes starch and protein damage but makes atta suitable for chapatti

Maida

The Indian subcontinent mostly uses two kinds of wheat flours: maida, which is made just from the endosperm, and atta, which includes a little bit of the bran. This is unlike the more 'wheaty' parts of the world—the Middle East, Europe and North America—where there is a cornucopia of variations based on the variety of wheat, how much of the bran is used to make the flour and how finely it has been ground. Of late, because urban Indians seem to have rediscovered millets, there has been an explosion of both gluten-free flours and wheat flours 'enriched' with millet flours. But, for now, we shall focus on wheat and discuss gluten-free breads in Chapter 7.

The milling process (in a *chakki*, which is a set of two millstones used to grind grain into flour) used to make atta causes a fair bit of damage to the proteins and starches in the flour, which makes atta not an ideal flour to bake leavened bread. A loaf of bread baked with atta tends to be dense and crumbly, and not soft and airy like it is if you use the whole wheat flour available outside India. This is also why leavened breads, such as naan and kulcha, tend to use maida, which is not made using the stone-grinding process and, thus, has better gluten development when leavened. If you want to make a whole wheat loaf of bread in India, your best bet is to use 70 per cent maida and 30 per cent atta for the best results.

Here is what happens when you add water to atta or maida. There are two proteins in wheat—glutenin and gliadin—that form a stretchy and elastic structure called gluten, which traps air to create give your finished bread a light and airy texture. Maida forms stronger gluten structures than chakki-ground atta, which is why chapattis made of maida are chewier than those made using atta. Gluten formation in a chapatti is focused on creating a soft, yet not overly chewy, superstructure. But in a loaf of bread, gluten formation is focused on making a strong structure that is able to handle the expanding gas generated by the yeast in the dough, finally turning it into a crisp brown crust at high heat in the oven, thanks to the Maillard reaction.

The more water you use to knead your dough, the softer the final product will be. But, remember, sticky dough can be difficult to handle. As a general rule, chapatti dough made with 100 per cent hydration (100 g of water for 100 g of atta) hits the sweet spot for dough-handling ease and softness of the finished product. If you are new to chapatti-making, start from 80 per cent hydration and work your way up as you get comfortable. And because this is a common error, it's important to distinguish between ratios by weight and ratios by volume. Most people making chapattis tend to use the two-cups-of-atta-with-one-cup-of-water heuristic, and this is perfectly fine because two cups of atta weigh about the same as a cup of water. When you use cups as a measure, it's volume, and when you use grams, it's weight. So, 2:1 ratio by volume for flour:water is not the same as a 2:1 ratio by weight because water weighs twice as much as flour for the same volume.

The algorithm for the perfect, soft chapatti dough is:

1. Mix atta and water, and roughly bring it to a shaggy mix (no need to knead) till there are no dry bits of flour. Let it sit for 30 minutes. This triggers a process of autolysis where gluten formation starts in the presence of water. You can use slightly warm, but not boiling, water to increase gluten development. Boiling water will cook (gelatinize) the starches in the wheat, and that will leave less water for gluten development. Some methods do recommend using boiling water, but that will produce not just a soft chapatti but also an ultra-flaky one. Ultimately, it's a matter of personal preference. I tend to like my chapattis with some amount of chew. The beauty of

autolysis is that you don't need to knead the dough at all. The dough will literally knead itself.

2. After 30 minutes, work in some salt into the dough. We don't add the salt up front because salt tends to tighten the gluten network, and we don't want that during the early stage of gluten development. Just knead the dough mildly till it looks shiny and slick (the autolysis phase will help make this happen pretty quickly) and you are done! Think of all those instructions that ask you to knead the dough for 10 minutes. If you think the exercise will be useful for your deltoid, triceps and biceps, go ahead, but it is not really necessary.

The actual rolling out and cooking takes some practice and experience, and there really isn't any other trick to it, so just keep doing it till you get better. Try and use as little extra flour for dusting and preventing stickiness while rolling, because all that extra flour tends to burn on the tawa and flame, adding a burnt flavour to the finished product.

Chapter 7 will explore more Indian breads, both leavened ones like naans and kulchas, which use yeast or baking soda/powder as leavening agents to make the dough airy and soft, and gluten-free breads made using millet flours.

Science of Vegetables

A fruit is one of the marvels of nature, a beautiful object with perfectly finished flavours enticing animals to eat them, and in the process, transporting the seed far away to hopefully grow into another plant. Fruits are inspirations for chefs. The plant is the chef here—it has perfected, over millions of years of evolutionary tinkering, the art of creating a balance of flavours so perfect that no amount of cooking can improve it. For instance, a perfectly ripe mango.

While fruits are designed to be attractive and delicious, vegetables and herbs are not. In fact, they are designed to keep animals from eating them. They tend to be hard to digest, and often possess nasty-tasting molecules. It is a testament to the remarkable nature of human ingenuity that we have figured out how to dig up a potato from the ground and turn it into the most scrumptious aloo fry using the process of cooking. The strong flavour of herbs like mint and coriander repel most insects. The sulphurous compounds in mustard, onions and garlic irritate not just insects but also animals grazing and trying to grab a quick bite. But humans tame their pungency and transform them into amazing flavours by cooking them. As we will discover in Chapter 2, most strong flavours are the plants' defence mechanisms.

A fascinating thing to think about is why organic vegetables tend to taste better than non-organic produce. As we just learnt, flavour tends to be a function of how strong the plant's defence mechanisms against predatory munchers is. If a plant is exposed to more pests, it will use more of its resources to defend against them and, thus, be more flavourful for us. This is why non-organic produce, grown in sterile and pest-free environments, tends to be bigger and lacking in flavour. One of the mechanisms a plant uses to sense insectile threats is to detect a substance called chitin. Chitin makes up the cell walls of several fungi and insects that attack plants. So, when a plant detects chitin, it goes into ninja mode and invests more in its defence budget. We can use this behaviour to trick the plant into thinking that there are pests nearby, by mixing powdered crustacean shells (like shrimps, which are closely related to insects) into the soil the plants are growing in. While we want the flavour that comes from plants operating in DEFCON 1 mode, we don't actually want pests to eat the stuff we should be eating. The chitin in the soil does the trick. Some modern organic farms use this trick to produce delicious vegetables with a reduced risk of pest damage.

Plant cell walls are made of pectin, a super-tough material that takes a fair bit of heat to break down. This is why vegetables require a higher amount of heat to cook, while meat overcooks at high temperatures. Vegetables can be cooked in a variety of ways, and it all starts when you chop them. Mechanical damage to many vegetables kicks off several enzymatic reactions that start 'cooking' and transforming them. Understanding this process is crucial to getting the right flavours out of every ingredient. For example, how you choose to chop can determine how much cellular damage you do and, therefore, the extent of the chemical reactions that modify taste. For instance, using garlic cloves whole will taste very different from chopped garlic, while minced garlic is a different ball game entirely. Likewise, green leafy vegetables discolour and turn bitter when exposed to heat for long periods of time. This is why the best way to cook leaves is to blanch them in boiling water for 30 seconds and then to remove them away from heat. If you need to use them in salads, you can also stop any further cooking by plunging them into a bowl filled with ice water. The application of high heat for a short period deactivates an enzyme called polyphenol oxidase that typically robs the chlorophyll molecule (the one that gives leaves their bright green colour) of its precious magnesium atom, causing the leaves to discolour into an unappetizing dull green. You can then puree the blanched greens and use it as you like. It will both look bright green and not taste too bitter. In general, high temperatures (around 85°C) for short periods of time are the best way to cook most vegetables to optimal taste and texture.

Other enzymatic reactions cause peeled raw potatoes (and several other vegetables) to turn brown when exposed to air. This is called enzymatic browning, as we learnt earlier. When a fruit or vegetable is damaged, some enzymes swing into action and go into the if-you-are-going-to-cut-this-precious-product-of-my-plant's-hard-work, I-am-going-to-oxidize-it-and-turn-it-into-brown-mush mode. We still don't entirely know what the evolutionary purpose of turning a vegetable or fruit brown and mushy is, but recent research indicates that plants with more of this enzyme tend to be more pest-resistant than others. This is why we store chopped vegetables in a bowl of cold water after cutting them, preventing access to air. Squeezing some lime juice into the water also prevents oxidation.

Another useful rule to remember is to always use salted water when boiling vegetables. Salt will prevent the leaching out of flavour molecules and nutrients from your vegetables, and also accelerate their cooking times. In general, steaming vegetables is a better approach than boiling them, especially if you are looking to retain most of the vegetable's flavour and texture. It will take longer than boiling because, as we learnt earlier in this chapter, vapour is less dense than liquid water and, thus, takes more time to transfer heat. However, if the idea is to assault the vegetable with fifteen spices, ginger and garlic, then it doesn't matter much. But if you are making a minimalist dry dish with a vegetable, it's better to steam the vegetable to the right level of doneness, brown it in oil and then add spices, before quickly turning the stove off.

Science of Meat

Vegetables are tender and moist when the water they are in nears boiling point (100°C), as heat causes the plant cell walls to weaken and absorb water. On the other hand, perfectly cooked meat is tender and moist when moderate heat, well below boiling point, is applied, causing the proteins to bind loosely to each other while still being able to retain water. If you overheat meat, and this is ridiculously easy to do, it will become tough and dry, and the proteins will bind to each other really tightly, squeezing all the moisture out. You can see this in all proteins of animal origin, from paneer to eggs to chicken. If you apply anything other than mild heat to paneer, it will turn into rubber, as will eggs or poultry. Meat (and animal protein in general) is edible in a very narrow range of temperatures. Below 55°C it is uncooked and potentially harbours dangerous microbes, while above 65°C it becomes dry and chewy. And to make things more complicated, how long you need to cook meat will depend on which part of the animal is being cooked.

To understand this better, we first need to understand how animals are built. The parts of an animal that we generally eat are made up of three components:

1. Fat: This is typically deposited under the skin, and in the case of large mammals, in between muscles (called marbling). Fat by itself is tasteless, but it transports flavour and thus makes meat taste better.

2. Muscles: Their taste and cooking time depends on what part of the animal they come from. Muscles that are used regularly take longer to cook (like chicken legs), while parts of the animal that generally Netflix and chill tend to cook really fast (chicken breast).

3. Connective Tissues: They take more time to cook. Slow and low heat makes them transform into gelatin, which gives meat its succulence. The tricky thing is that this happens between 65°C and 70°C, at which temperature the muscle tissue starts to dry out, so the trick is to balance protein denaturation of muscle tissue and breakdown of connective tissue to get the perfect texture.

A combination of these three, and their ability to retain as much of the 70–85 per cent water that all animals are made of, determines how tender and moist they are when cooked.

So, now that we know how to keep meat moist and tender, the next question is: How do we impart additional flavours to meat? You'd be tempted to think of marination, since every other recipe tends to call for overnight marination in a spice mix that includes an acid like yoghurt or lime juice. Again, this is a classic case of looking at a delicious finished product and describing the wrong route to get there. Marinades, despite conventional wisdom, do not penetrate into the meat. At best, they coat the surface. When cooked and eaten, your mouth usually can't tell the difference between flavours that come from the surface and those that come from inside the meat.

What does work in getting flavours into the meat is a process called brining. Letting meat sit in a salt solution, into which you can add other flavouring ingredients as well, will result in the salt getting into the meat, which by itself significantly improves the flavour. More magically, the salt prevents the loss of water from the muscle tissues in the meat. This is rather counter-intuitive because when you add salt to vegetables, they tend to lose water, and we tend to assume that salt is a dehydrating agent. Yes, it is, but only for plant cells. For animals, including humans, salt helps retain moisture. Think about what you do when you are dehydrated. You drink water (to replace the lost water), sugar (for instant energy) and salt, which helps you retain the water you just drank.

So, should you stop marinating? No. It's still a decent way to get flavour to stick to the surface of meat before you cook it. And since it's not a good idea to cook meat for too long anyway, it's better to get the flavour on to the surface well before you apply heat. But brining is absolute magic. It will transform the taste of meat. Given urban India's general tendency to never take risks when it comes to meat, most dishes made at home tend to be overcooked. It is the sophistication of the gravy and flavouring that compensates for what is usually a rather dry and chewy piece of meat. But brining will give you the best of both worlds. It will help you retain moisture inside the meat while cooking it enough to kill microbes and inject flavour all over while you are at it.

Another common frustration with meat is defrosting. You have to store it in the freezer to prevent bacterial growth, but we all know how frustratingly long it takes for meat to thaw fully. Leaving it at room temperature for too long is a welcome-to-the-free-all-you-can-eat-buffet neon signboard for travelling bacteria. Food scientists recommend thawing at 4–5°C, which is just above freezing point. This means you need to move the meat from the freezer to the regular part of your fridge and leave it there for several hours for it to thaw fully. So, applying what we now know about how meat cooks, keeping it in a water bath at 39°C, will defrost your meat in 10–15 minutes without cooking it. Protein starts to denature around 50°C, so the idea is to get water as warm as you can while ensuring that it is circulated constantly to ensure even transfer to heat. A sous vide device will do this, but if you do not have one, heat water in a microwave, at a low setting for about a minute. Then use a temperature probe and wait till the water reaches around 40°C. Put your meat in it and keep stirring once in a while. Tap water in India is regularly in the 40–45°C range during summer, so you can use that, but make sure you use a lid (or cling wrap) to reduce exposure to air.

Science of Eggs

The magic of an egg is that an entire living thing can be created from its ingredients, with one large caveat though. An egg will hatch only if a rooster has had the opportunity to do some *jalsa* (Chennai Tamil/ Urdu slang for having a good time with a partner) with the egg-laying hen. And if you didn't know this already, there is a multi-billion dollar global industry dedicated to preventing roosters from doing so, which allows us to enjoy the hens' daily offering instead of the egg giving up its nutrients to a hungry baby chick. In fact, the very notion of an egg being considered non-vegetarian in India, while milk squeezed out from the udders of a post-partum cow is considered vegetarian, is predicated entirely on the belief that the egg is a prospective chick only if you let it sit around and wait for it to hatch. Unfortunately, that logic is a bit like claiming that the jackfruit tree in your neighbourhood is a potential study table in the complete absence of carpenters. The egg in your fridge has about as much chance of hatching as a chick as that jackfruit tree has of metamorphosing into an imposing study table.

Unlike meat, eggs have a tremendous variety of proteins, and each of them behaves differently when heat is applied. This makes an egg the single most versatile ingredient in the kitchen. It can be scrambled into a soft and unctuous bhurji, it can be turned into an omelette that has a crisp outside and a melt-in-the-mouth texture inside, and it can be emulsified into mayonnaise with some fat. And if all of that is too complex, it can simply be boiled to every possible variation of texture— from gooey, soft-boiled eggs to firm, hard-boiled eggs. I haven't even talked about the use of eggs as a binding agent for bread crumbs to make the crispest mutton cutlets, or its ability to make your naans soft when added to dough. And have I talked about desserts yet? You get the point, I am sure. If eggs didn't exist, food would be rather one-dimensional.

Let's first see what happens when you drop an egg into boiling water. The protein coils in the egg whites unfold and start bonding with themselves. As you continue to apply heat, the translucent whites turn opaque, and if you cook it any further, hydrogen sulphide (H_2S) gas will be released, resulting in the quintessential eggy smell no one likes. Fun fact: H_2S is also the primary smell of black salt (kala namak), which has a tiny bit of H_2S in it. While the whites are setting, the yolk proteins get crumbly. If you overcook the egg, the iron in the yolk reacts with the H_2S from the whites to produce ferrous sulphide, which is that unpleasant green deposit you see between the yolk and white in an overcooked egg. And if you have the problem of the egg sticking to the shell after boiling, making it painful to peel, try adding a pinch of baking soda to the boiling water. This will increase the pH level of the water, which will keep the egg from sticking to the shell.

As with meat, the best way to cook an egg is to use low heat. If you are scrambling an egg or making an omelette, the trick to getting the softest and fluffiest results is to salt the broken egg at least 15 minutes before you cook it. The salt will uncoil the proteins in the egg before they have the chance to set rapidly when heat is applied. This allows for a softer texture in your omelette or bhurji.

Science of Fat

Fats have a PR problem. For starters, the word has no positive connotations. It suggests greed and corpulence, and when deposited on our body parts, arrogance (in Tamil). We have spent decades focusing *more* on how to eat *less* of the wrong kind of fat, and *less* on eating the kind that is both absolutely essential to life and critical to the flavour of the food we eat. At 9 calories per gram, of course, we must pay attention to how much fat we consume, but there is an inordinate asymmetry in the obsession to reduce fats in our diet compared to the real villain, carbohydrates. But this is not a book about nutrition, so we will proceed with the broad understanding that some kinds of fat are best kept to a minimum, while others must be embraced to truly unlock culinary greatness.

*Representative image only

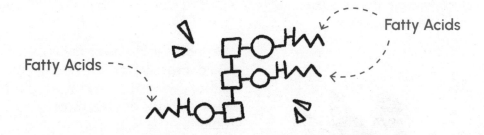

How scientists represent fat molecules

Cooking oils are distinguished on the basis of

Whether the fatty acids are long or short	Whether the fatty acids are saturated or unsaturated	Number and location of points of unsaturation

Your diet should ideally include 1 tablespoon of fat per day (~500 ml per month)

Fats should comprise 20%-30% of calories in your diet

You can choose any oil, but make sure it's not the only kind you use. Instead use multiple for balance

Refined oils for deep frying, Cold-pressed for low heat situations, saturated fats - less that 10% of your total calories

58

To understand how fats work, we must understand what they are. A molecule of fat is like a flagpole with three flags. The flagpole is a small molecule called glycerol and the three flags are fatty acids. While they are called 'acids', they are pretty weak. If you fill your car battery with them, it won't help your vehicle start. There are many kinds of fatty acids, some that are long chains of carbon atoms and some that are short, more prone to flying off the flagpole, a behaviour that is crucial to understanding how fats go rancid, but we are getting ahead of ourselves here. Fatty acids, both long and short ones, have two kinds of bonds between the carbon bonds—a single bond or a double bond. When there is a double bond, there are fewer hydrogen atoms, as the carbon bonds once more with carbon instead of with the hydrogen in the fatty acid. When all the bonds in the fatty acid are single bonds, the fatty acid is saturated (with Hydrogen). When one of the bonds is a double bond, it's a monounsaturated fatty acid. When there are more than one carbon double bonds, it's a polyunsaturated fatty acid. Each fat molecule could have any combination of three saturated, monounsaturated and polyunsaturated fatty acids.

So, what does all of this mean in the kitchen, beyond the fact that you can now read the labels on the oils you buy and mentally picture flagpoles and flags of fatty acids waving about. Any oil that has a higher percentage of saturated fatty acids is more likely to be solid at room temperature. Think of butter, ghee, coconut oil and animal fat (lard). Any oil that has more unsaturated fatty acids is likely to be liquid at room temperature. Think of groundnut, mustard or sunflower oil. Because solids are easier to transport, the fat industry has, over the years, tried to convert cheaper sources of oil (typically from plants) that tend to have more unsaturated fatty acids into solids. This is done by forcing hydrogen into those fatty acids and turning the double carbon bonds into single bonds through a process that we should be familiar with, thanks to the label on Dalda—hydrogenation. So, while the palm oil from which Dalda is made is liquid, Dalda itself is solid at room temperature. Saturated fats also have a longer shelf life than unsaturated fats.

This brings us to omega-3 and omega-6 fatty acids. Because chemists tend to be insufferable geeks, they describe unsaturated fatty acids in terms of where the carbon double bond is in the long chain. If it is three atoms from the end of the chain (Omega is the last letter in the Greek alphabet), it's called Omega minus 3. So, now you know. This family of fatty acids has been shown to have some positive effects on cardiovascular health, thus the tendency to overuse the term in advertising.

Fats transport flavour and there is now some consensus that we have taste buds that can detect fat. This is why when there is too much of it, we find the food greasy. Most flavour molecules in spices dissolve only in fat, not in water. This may be the reason why more or less every single dish made in India starts with whole spices and flavouring ingredients being added to hot oil. That base of flavour is what defines the dish. Spices added to water, in contrast, lose most of their aroma to the air because flavour molecules do not dissolve in water.

Fats, unlike water, have higher boiling points. In fact, in the context of fats, it's called the smoke point, because while water boils away into vapour, oils usually catch fire and burn when heated well beyond their smoke points. This is also why deep-frying is generally a dangerous activity in the home kitchen, as the best results are achieved very close to the smoke points of the oils used. The choice of fat used for deep frying is important. Given that there is now a tendency to buy unrefined or 'virgin' oils for use at home, please remember that the smoke points of unrefined oils are almost always lower than the temperature you need for effective deep-frying. And, by the way, when you heat oil, any fancy aroma that an unheated expensive oil has all but disappears. So, buy expensive oils only as finishing oils, not cooking oils.

Fat	Smoke Point	Use for frying?
Butter	150°C	No
Ghee	250°C	Yes
Refined coconut oil	232°C	Yes
Virgin coconut oil	177°C	No
Refined olive oil	200°C	Yes
Extra virgin olive oil	160°C	No
Palm oil	235°C	Yes
Refined groundnut oil	227°C	Yes
Unrefined groundnut oil	160°C	No
Rice bran oil	232°C	Yes
Safflower oil	266°C	Yes
Refined sesame oil	232°C	Yes
Unrefined sesame oil	177°C	No
Sunflower oil	252°C	Yes
Vegetable oil (blended)	220°C	Yes

In general, it's a good idea to use a virgin or unrefined oil for day-to-day cooking, and a refined oil with a high smoke point for deep-frying, which is not something you are going to do every day.

It's also important to understand how oils go rancid. In general, oils with more unsaturated fatty acids are more likely to go rancid. Remember the glycerol flagpole with the three fatty acid flags attached to them? As long as the fatty acids are attached to the glycerol, things are fine, but if they happen to escape the flagpole and loiter about on their own, things get nasty. It turns out that individual fatty acids are extremely noxious, and nasty-smelling and tasting, molecules that somehow, when attached in threes to glycerol turn into edible oils crucial in the kitchen. So, how

61

do rogue fatty acids break free of the glycerol? Light, water and oxygen. Light tends to act as a catalyst for the rebellious behaviour of fatty acids, and when you deep-fry something in hot oil, you are essentially dropping something with a ton of water into fat. Hot water, too, has a tendency to break some of these bonds, which is why oil used for frying has a characteristic 'used oil' smell. This is essentially early-stage rancidity. Use it a few more times and it will become unpalatable. What oxygen (essentially, exposure to air) does is that it goes straight for those double-carbon bonds in unsaturated fatty acids and cleaves out funky-smelling molecules called aldehydes and ketones. Fun fact: It is in very few situations, such as funky-smelling European cheeses, that these strong odours are desirable, but definitely not when making chicken jalfrezi. And, by the way, rancid oil only smells and tastes bad. Consuming it has no known serious, negative health effects.

So, as the labels will tell you, store your oils in a cool, dark place, and keep an airtight lid on at all times. Pay close attention to the smoke point before determining what to use it for.

2 Science of Spice and Flavour

The plural of spouse is spice.

—*Christopher Morley*

Consider coriander. Its name comes from *koris*, the Greek word for bedbugs, because the ancient Greeks thought that the seeds smelt like the insect. It is, of course, unfortunate because, if anything, we must say that the bedbug smells like the spice and not the other way around. Humankind's association with those critters is likely more recent than our association with this versatile spice, which finds its way into almost every Indian dish in every imaginable form: leaf, stalk, root and seed. But there is a sizeable section of the population that has a visceral aversion to coriander (4–14 per cent depending on their ancestry) in its leaf form. It turns out that it's not an irrational personal choice but genetics. Coriander's flavour molecules are a family of compounds called aldehydes, and the ones present in coriander are also found in soap. Some people have a gene (or a combination of genes, we don't know fully) that makes their taste buds specifically sensitive to aldehydes, so eating coriander leaves strongly reminds them of soap.

But what's interesting is that when you crush coriander leaves, or grind them into a paste, an enzymatic reaction breaks down these soapy aldehydes, which is why people who can't tolerate the leaves as a garnish don't mind coriander chutney in their chaats. Beyond those sensitive to the flavour of coriander, Professor Linda Bartoshuk from the University of Florida has coined the expression '*supertaster*' to describe people with a genetic predisposition that enables them to detect bitter tastes more strongly than others. Using a strongly bitter compound called propylthiouracil, the research tells us that about 25 per cent of the population is super-sensitive to the bitterness of this compound and cannot tolerate it, while another 25 per cent cannot detect it at all! The rest of the population has a spectrum of sensitivity that ranges from low to medium. Supertasters are also more likely to be chefs and food critics. And women.

So, this sordid tale of bedbugs and genes that make you taste soap more strongly than others brings us to the subject of this chapter—how to think about flavours in your dishes, so that you use the right combination of spices, herbs and other high-flavour ingredients. If you understand where flavours come from, how to extract exactly as much as you want from your ingredients (maximum is not always a good idea since it can overpower a dish) and how to combine flavours in ways that behave more like Dhoni's XI, circa 2011, as opposed to Sachin's XI, circa 1997, it will help you take the single biggest leap in day-to-day cooking

Taste and Flavour Perception

When you bite into pani puri, also called golgappa or phuchka, the dominant aroma of cumin and mint, the sour tang of the amchoor, the heat of the green chillies and the satisfying crunch of the puri contrasting with the soft and creamy texture of the filling (which varies by region) is what makes up the entirety of the pani puri experience. Flavour is a combination of taste, smell, mouthfeel, and to a smaller extent, sound and visual experiences. And despite the fact that 80 per cent of flavour perception happens in the nose, we tend to associate the tongue as being the Watson and Crick of flavour to the nose's Rosalind Franklin.

Taste

Let's start with taste. Our tongues have taste buds that can detect five primary tastes. Contemporary research tells us that the old picture of the map of the tongue, with distinct taste perception areas, is mostly wrong, and that while the tip of your tongue does have more taste buds dedicated to detecting sweet and salty flavours, it doesn't mean that they cannot detect sourness or bitterness entirely. Taste bud specialization is reasonably well distributed. Sweetness and saltiness are detected rather quickly, while bitterness, which is mostly detected at the back of the tongue, takes a bit longer and tends to linger in the mouth. This explains the expression 'bitter aftertaste'. Sourness tends to be detected more strongly on the sides of the tongue. What kind of molecules tend to be sweet? Sugars, aldehydes, alcohols and certain amino acids taste sweet to varying degrees. Acids, such as tamarind, vinegar, yoghurt, juices of citrus fruits and many other organic acids in fruit juices taste sour and tart. Saltiness comes from, well, salts, of which sodium chloride tastes the saltiest, while other salts exhibit varying degrees. Bitter tastes come from substances called alkaloids, such as caffeine (in coffee), theobromine (in chocolate), quinine (in tonic water), and so on. Our ability to detect bitterness comes from the need to identify poisons before we ingest them. Pure caffeine is deadly and poisonous, although a tiny amount as part of our morning coffee is typically safe. Umami is the fifth taste that has recently been added to this list. It is the savoury, lingering, meaty taste that comes from the presence of salts of a specific amino acid called glutamic acid. Food that is rich in glutamates has an intense, savoury and

So How Do We Really Taste Things?

Gustatory
Receptor cells

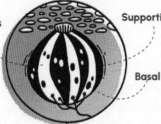

Supporting Cells

Basal

In each taste bud there are roughly 50 gustatory receptor cells and basal and supporting cells. They are contained in the papillae.

On each gustatory cell there is a gustatory hair which extends just outside the taste bud through an opening called a taste pore.

Gustatory Hair

Taste Pore

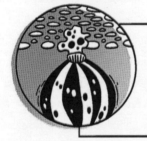

When you put something in your mouth those molecules mix with saliva. This mixture enters through the taste pore, interacts with the gustatory hairs and stimulates the sensation of taste.

When the sensation is stimulated, it activates the gustatory impulse, receptor cells then synapse with neurons and pass electrical impulses to the gustatory area of your brain.

Gustatory

Your brain then interprets the sensation as taste.

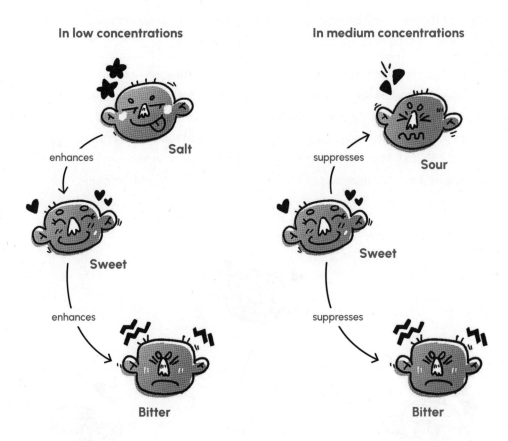

In low concentrations

Salt

enhances

Sweet

enhances

Bitter

In medium concentrations

Sour

suppresses

Sweet

suppresses

Bitter

Relative sweetness of sugars and sweeteners

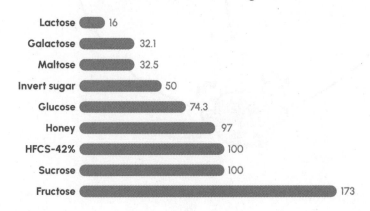

Lactose	16
Galactose	32.1
Maltose	32.5
Invert sugar	50
Glucose	74.3
Honey	97
HFCS-42%	100
Sucrose	100
Fructose	173

lingering flavour that feels very satisfying. Umami-rich foods do not need to be overly spiced or salted because of this lingering effect.

The perception of taste is also dependent on the concentration of the substance responsible for the taste, and different individuals have different thresholds for perceiving tastes. For instance, a lot of Indians will need salt at least at 1–1.5 per cent concentration by weight of the dish to taste acceptably salted. Many in the West will find that too salty, as they can perceptibly detect saltiness at much lower concentrations. In addition to

In low concentrations

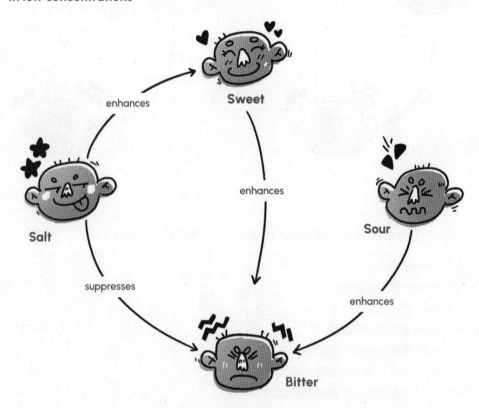

In medium concentrations

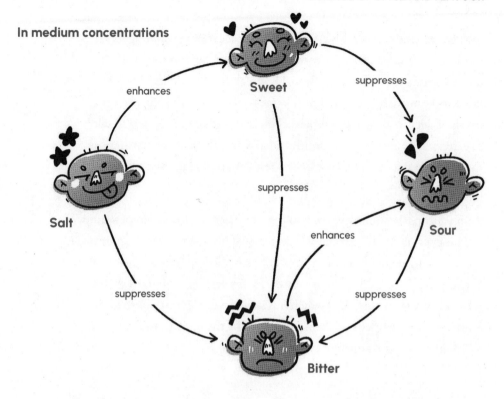

traditional culinary habits, genetics also play a role, as we saw in the case of coriander leaves. People with low thresholds for detecting the soapy, bitter taste of aldehydes will find coriander leaves unpalatable.

Fascinatingly, food scientists (and grandmothers and mothers) have figured out that there is a sub-threshold level of taste, which while not being individually detectable, can amplify or mute other tastes. For example, a tiny pinch of salt in your kheer can make it taste more intensely flavourful without being perceptibly salty, which no one would want. This is also why jaggery tends to be preferred when making Indian desserts, because it naturally contains a bit of salt. Likewise, a tiny pinch of sugar can mute saltiness without tasting perceptibly sweet. This is why a pinch of sugar is a good idea in any dish, because it balances saltiness. In fact, restaurants tend to take this effect to its logical extreme—adding lots of sugar allows you to add lots of salt to your dish. This combined effect is like turning the volume knob to 11, which is why restaurant food tastes more intensely flavoured than home-cooked food. And finally, a tiny pinch of salt can also mute sourness. This is why it's common to add salt to extremely sour yoghurt in south India, where the local climate tends to supply crystal meth and cocaine to fermentation reactions, to make it palatable.

Temperature also impacts how you perceive taste. At lower temperatures, our tongues' ability to detect tastes decreases. This is why melted ice cream tastes cloyingly sweet. It turns out that taste buds operate at their peak between 20°C and 30°C. This is why coffee is tolerable at 50–60°C, which is usually the temperature at which it is served, while it tastes bitter once it gets to room temperature, which tends to be the temperature range in which our taste buds operate at their peak.

Aroma

Now, let's talk about the most impressive, yet underrated, arsenal in our flavour-detection apparatus—the olfactory receptors in our noses. When you pick up that succulent piece of tangdi kebab, marinated in a ton of spices, tenderized by yoghurt and seasoned with salt, and bring it up to your mouth, the aroma molecules that escape the kebab enter your nose and hit these receptors, which then send out a message to the olfactory cortex that can recognize more than 10,000 unique flavours. But we are just getting started. You then bite into the kebab. Since saliva is mostly water, your tongue can only detect tastes coming from water-soluble flavour molecules. The gnashing of your teeth causes the release of more volatile flavour molecules that travel through the back of your mouth, into the nasal cavity, and hit the same receptors as you breathe out. What

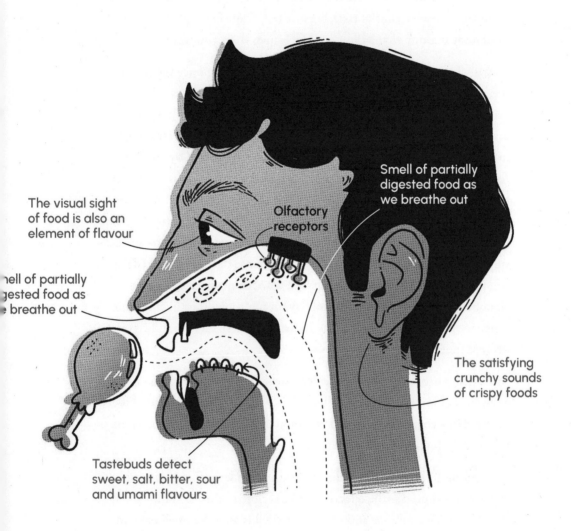

Smell of partially
digested food as
we breathe out

The visual sight
of food is also an
element of flavour

Olfactory
receptors

ell of partially
ested food as
breathe out

The satisfying
crunchy sounds
of crispy foods

Tastebuds detect
sweet, salt, bitter, sour
and umami flavours

you smell before you eat is called *orthonasal* olfaction, and what you smell as a result of a ton of volatile flavour molecules hitting those receptors as you breathe out is *retronasal* olfaction. This, in my opinion, is the single largest contributor to taste, because the act of chewing releases the maximum amount of volatile aroma molecules from your food.

This is why many foods that smell funny taste amazing once they are in the mouth. A good example is cheese. Many older Indians find the funky smell of cheese off-putting. This is because what you smell before you eat the cheese is the overwhelming smell of the ketones and aldehydes in it. Once you chew it, the complex molecules generated by the slow fermentation process start to hit your nose retronasally, and that's when you go 'ah, cheese'. Flame-roasted brinjal is another example. It smells acrid and metallic but is delicious once you actually chew it. Fish is another interesting example. Most people not used to fish will first taste and smell a sulphurous molecule, which makes up the intense flavour of cooked fish and can be rather off-putting for anyone who has never eaten fish before. But once you chew, a ton of umami flavour molecules start hitting your tongue, and that is what makes any Indian fish curry such an addictive dish for those who are used to fish.

More than taste, aroma is highly culture-specific. For instance, the smell of roasted cumin powder smells like sweaty feet to people not used to cumin. Now go smell that bottle of cumin powder and silently curse me for putting this idea in your head.

Malodorous feet and cumin aside, smell is the only sense that goes straight to the brain's cortex—the olfactory nerve is close to the part of the brain that deals with emotions and memory, which is why the smell of food evokes nostalgia and memories, and also why no Michelin star chef can compete with your grandmother's dal. After all, it is not objective taste and aroma that matters but the fond memories associated with it that come rushing back when you eat a good home-cooked dal.

Mouthfeel

We can all agree that, scientifically speaking, a dense idli and an airy idli are chemically the same and should theoretically taste the same. But they obviously don't. In addition to taste and aroma, our flavour detection apparatus has the ability to distinguish textures. According to food scientist Alina Szczesniak (pronounced Chesniak), we can detect:

1. Cohesiveness

2. Density and heaviness

3. Dryness and wetness

4. Fracturability (as applicable to potato chips and pani puri shells)

5. Graininess

6. Gumminess

7. Hardness

8. Mouth coating

9. Roughness

10. Slipperiness

11. Smoothness

12. Uniformity

13. Viscosity

So, it's not just taste and aroma. If the texture is wrong, and wrong here essentially has to do with a learned preference for a specific, associated texture with that taste and aroma, we will find it off-putting. For instance, an idli can have that perfectly mild, sour taste and the pleasantly nutty aroma of fermented urad dal and rice, but if its texture is not airy, we are likely to leave a bad rating for it on Zómato.

Sound and Sight

Remember that perfectly crisp potato chip? It turns out that the satisfying sound of the crunch is an integral part of the flavour experience. Our preference for crispy and crunchy sounds in food comes from the evolutionary preference for fresh plant products. Crisp and crunchy fruits, and stems and nuts, indicate freshness and, thus, nutritiousness. A rather fascinating aside here: The sound of a crisp potato chip being bitten into is incredibly loud in terms of decibels, and a good amount of the sound is in the ultrasonic range that we typically cannot hear because we are not bats. But we can feel ultrasound because it travels through bone more efficiently than regular sound, which is why hearing-challenged people can also feel the crispness of a pani puri, even if they don't necessarily hear the audible crunch. We can, it turns out, feel a good pani puri in our bones.

Science of Salt

Chemically speaking, a salt is a cosy arrangement between a positively charged atom or molecule and a negatively charged one. But culinarily speaking, salt refers to a specific kind—the most ubiquitous salt on the planet—sodium chloride. Our tongues typically detect sodium in food and declare it to be salty. Pretty much any salt of sodium, from sodium bicarbonate (baking soda) to the infamous MSG will taste salty to varying degrees, with sodium chloride being the saltiest of them all. Potassium chloride, for instance, does not taste salty and is used to make low-sodium salt, a dubious product that is a mixture of sodium chloride and potassium chloride, targeted at people looking to reduce their salt intake. The mixture tastes less salty because there is less sodium by proportion. Potassium chloride is also used in the infamous 'three-drug cocktail' that makes up the lethal injection used for prisoners on death row in the United States. At high-enough concentrations, it can stop the heart. On that macabre note, let us return to our friendlier star of the show, sodium chloride.

It is a strange and mysterious substance. This simple molecule brings food to life, and yet it is the only substance we eat that did not originate in a living thing. It draws out flavour like a magician pulls a rabbit out of his hat (FYI, this metaphor works because the rabbit, like flavour, was always there in the hat but won't come out without the magician pulling it out). It increases aroma, balances sweetness and sourness and reduces bitterness.

Since our saliva contains 0.4 per cent salt by concentration, any food with less than 0.4 per cent salt by weight will taste unseasoned. So, 0.5 per cent to 1 per cent by weight is a sensible starting point for estimating how much salt you need in your dish. It may not be enough to your taste, because salt tolerance varies by cultures and individuals, and it's not uncommon for folks in the subcontinent to tolerate salt levels from 1.5– 2 per cent by weight, but as a general good habit, getting acclimatized to 1–1.5 per cent is a good way to enjoy the more subtle flavours in food instead of simply being assaulted by salt, with the added bonus of keeping your heart in relative shipshape.

The most commonly used salt in India tends to be iodized salt, which is sodium chloride mixed with tiny amounts of potassium or sodium iodide, a practice that started in the late 1950s to address the problem of goitre, a serious thyroid malfunction caused by the deficiency of iodine. There is, however, one problem. Iodide salts tend to break down at high temperatures and lend an acrid metallic taste to food. You will not notice this when making gravies because the temperature is not going to exceed 100°C in those cases, but you will notice it when you deep-fry or bake food in an oven. This is why it is recommended to use non-iodized salt when baking or deep-frying food. By the way, the processed food industry usually never uses iodized salt for this reason. While I do not want to underplay the seriousness of iodine deficiencies, it's not too hard to get your iodine from other dietary sources, such as dairy products, seafood and several fruits and vegetables, and use non-iodized salt in your kitchen. At the very least, use iodized salt when making gravies and non-iodized salt when deep-frying.

This brings us to the kinds of salt typically used in Indian cuisine:

1. Iodized table salt: This is the most commonly used salt. It can leave a metallic taste if used at high temperatures, like for deep-frying or baking.

2. Sea salt: It is generally used as a finishing salt. It has a large crystal size and provides a burst of flavour to food, as opposed to dissolving evenly.

3. Pink salt: Rather popular of late, it comes in multiple crystal sizes and tends to have a lot of other minerals in addition to sodium chloride. The health claims are entirely dubious and the saltiness levels are lower than that of regular table salt. Use 1.25 teaspoons of this for every teaspoon of table salt.

4. Rock salt (sendha namak): This is the traditional kind of salt used in India and generally tends to be large in crystal size. Use 1.25 teaspoons of this for every teaspoon of table salt.

5. Black salt: This is sodium chloride mixed with several sulphurous salts of iron and sodium, and trace amounts of hydrogen sulphide, which lends it its characteristic pungent

flavour. Use 1.5 teaspoons of this for every teaspoon of table salt to achieve a similar level of saltiness, although it's almost always used along with regular salt and never as a replacement.

A rather common question people have is how to fix a dish that has been over-salted. Allow me to use state-of-the-art science to finally give you the highly anticipated answer.

You can't.

Adding potatoes, balls of rice or dough don't really reduce salt concentration. All they do is absorb the gravy and leave you with less gravy overall. You can add sugar or an acid (like lime juice) to change the perception of saltiness, but that won't move the needle by much. What does work is adding more unsalted stock (or just plain water), but that will dilute your dish.

When cooking, an important salt behaviour to keep in mind is that while it dehydrates vegetables, it prevents moisture loss in meats. Adding salt when cooking vegetables will cook them faster and adding salt to raw vegetables will cause them to lose water. But if you let meat sit in a salt solution for a few hours before cooking, it will remain tender and moist after cooking. This magic trick, which we discussed briefly in the previous chapter on the science of meat, is absolutely key to preventing meat from drying out during the cooking process.

Here's a quick cheat sheet on the rules of salt:

1.   Salt balances sweetness by elevating other flavours in desserts. Take your dessert game to the next level by always adding a pinch of salt when making something like payasam or kheer. You will not be disappointed.

2.   Salt mutes sourness. This is a trick commonly used to salvage overly sour yoghurt by serving it with a pinch of salt.

3.   Salt mutes bitterness. This tip is very useful when cooking green leafy vegetables or bitter gourd.

4.   Salt dehydrates plant material. This is an excellent trick to improve the flavour of salad ingredients.

5.   Salt helps meat retain moisture. You must always consider brining meat before marinating or cooking it.

Science of Sugar

Sugar is among the most misunderstood things in the Indian culinary landscape. This is surprising because we produce, sell and add more sugar to our food than any other people on the planet. This is also despite the fact that the very idea of extracting sucrose from the sugarcane plant was originally Indian. The word 'sugar' and its equivalents in every language, from Persian to Arabic to European languages, follow the path that sugar itself took from its origins in what is today Bengal. It is derived from *sharkara* in Sanskrit. Fun fact: Even the word 'jaggery' comes from the Portuguese *jagara* that comes from the Malayalam *sakkara*, which again goes back to the Sanskrit sharkara.

Originally used to make bitter medicine palatable, sugar is, chemically speaking, a family of molecules that are water-soluble carbohydrates. Incidentally, not all sugars taste sweet. Sucrose is the one that is most familiar because it makes up the crystalline sugar we use every single day. Sucrose by itself is made up of two other sugars—glucose and fructose— that got together, shook hands, agreed to lose a water molecule and bonded together.

Glucose and fructose taste sweet individually too. The former is important because it is the single most important source of energy for all living things on the planet. All carbohydrates are ultimately broken down to glucose, the simplest possible sugar. This is why when your body is not functioning normally, and your digestive system is not able to take complex foods and turn them into glucose, hospitals stick a needle into your arm and pump glucose straight into your blood, bypassing the state-of-the-art organic factory that is your digestive tract. The other sugar, fructose, is largely found in fruits, which is why they taste sweet. Milk has lactose, which does not taste sweet and is a tricky sugar because most adult humans lose the ability to digest it (meaning, convert it into glucose). This is why adults mostly cannot consume large amounts of milk beyond the tiny amount in their coffees and teas, and the occasional kheer or payasam.

All starches, which are basically large complex molecules made up of simpler sugar molecules, are ultimately turned into glucose by the body. This is why when you chew on potatoes for long enough, the enzymes in your saliva will turn the starches into glucose, making it taste sweet.

So, that's about as much useful sugar chemistry theory one needs to know before jumping into the kitchen. The most common sweeteners in the Indian kitchen, in order, are:

1. Plain, crystalline white (or brown) sugar: White sugar is near 100 per cent sucrose. Brown sugar is white sugar with some molasses added back (the syrupy stuff that is left behind when refining sugarcane into refined sugar). This is the sweetest-tasting sugar.

2. Jaggery (gur): Jaggery is the unrefined mix of molasses (which is mostly glucose and fructose) and sucrose. It tends to be about 50 per cent sucrose, while the rest is mostly glucose, fructose and moisture. It has a slightly less sweet taste than sucrose but more depth of flavour.

3. Honey: This is mostly fructose and glucose, and has a very complex depth of flavour compared to plain sugar, or even jaggery. But the complex flavours are heat-sensitive, so avoid adding honey earlier in the cooking process.

Sugar needs to be at least 0.75 per cent by weight in your dish for it to register as sweet. But like salt, sugar can magically improve your dish even without being perceptibly sweet. In general, a pinch of sugar will improve any dish.

Here are some simple rules for sweetness as a taste:

1. Sweet mutes saltiness up to a point, and also mutes sourness and bitterness. You can use it to balance these flavours.

2. Sweetness adds depth to other flavours, such as spices. When you bite into a cardamom, you will smell it, but it will taste bitter. When you bite into cardamom with a pinch of sugar, the aroma and taste of cardamom will seem stronger.

Science of Heat

For science, let's do a small experiment. Go pick up a green chilli and bite into it. Now, let me explain what is happening to you. First, you taste the mildly citrusy and floral notes of the outer flesh of the chilli and the slight bitterness of the seeds. At some point, you will bite through the placenta, the white structure that holds the seeds to the outer flesh, after which a family of molecules known as capsaicinoids get down to work like the first batch of muscled prisoners exiting a prison they've just set fire to. As part of an evolutionary strategy to shield you from acts of wanton stupidity, your mouth has TRPV1 receptors, which trigger panic bells when a few things happen. One of them is when you bite into a hot samosa that you think has cooled down. The second is when you imbibe something with very low pH levels, essentially highly acidic things.

The TRPV1 receptors detect high temperatures and strong acids in the mouth, and here is where the genius of the chilli plant comes into play. As part of its evolutionary strategy to prevent being munched on by goats and cows, its neat little biochemical trick is the production of a family of molecules that snugly fit into this receptor and, like the proverbial boy who cried wolf, turn it on.

The receptors, as per standard operating procedure, start a chain of communication that notifies the brain about the mouth literally being on fire. The headquarter takes emergency action, like unleashing the experience of pain to remind you to not put hot things into your mouth, or rushing more blood to your face and increasing perspiration to cool you down. So, at this point, I take it that you must be in pain, sweating, face flushed and seeking some water. I'm afraid water won't be of much help. Capsaicin is not water-soluble, it only dissolves in fat or alcohol. A glass of milk, a spoonful of sugar or honey, or some wine, will be more efficient in putting out this (illusory) fire in your mouth.

Chillies aren't the only plants to have figured out this neat little trick to prevent animals from eating them. Mustard (allyl isothiocyanate), ginger (gingerol) and black pepper (piperine) also produce molecules that can trigger these receptors. By the way, you can desensitize this receptor by continuing to eat more chillies. Eventually, like the villagers in the story about the boy who cried wolf, they will not ring alarm bells every single time.

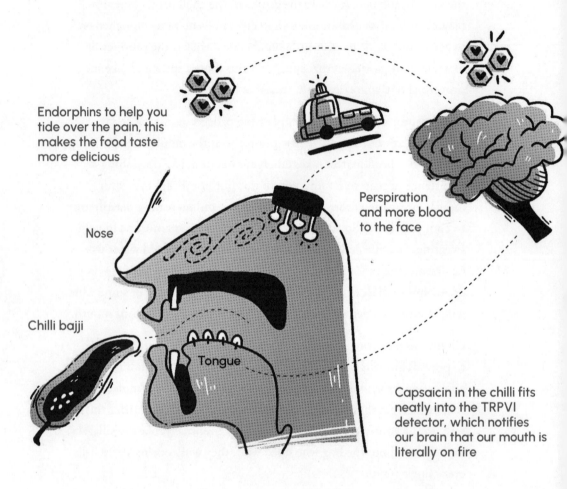

Endorphins to help you tide over the pain, this makes the food taste more delicious

Nose

Chilli bajji

Tongue

Perspiration and more blood to the face

Capsaicin in the chilli fits neatly into the TRPVI detector, which notifies our brain that our mouth is literally on fire

But why do we love chillies so much? For a plant that was unknown to the subcontinent till the Portuguese introduced it to us, after the Spanish discovered it in Mexico, chillies have come to define Indian food more than any other flavour. To understand this, we need to head back to the scene where the TRPV1 receptors panicked the brain into thinking that the mouth was on fire after being fooled by capsaicin. Once the brain deals with this panic, it has an automatic tendency to release endorphins, because sustained pain tends to incapacitate the body. Picture a Paleolithic man who has just been scratched by a wounded sabre-toothed tiger. It's a deep scratch and he is in a lot of pain, but if doesn't get on his legs and run away, the tiger is likely to make a meal of him. So, evolution has designed a mechanism where pain is usually followed by the release of endorphins. This convinces the opioid receptors in the brain to reduce the perception of pain, allowing the Stone Age chaps to run away from large cats even when injured. In simpler terms, the pain of eating chillies is also pleasurable, and since the capsaicin is only creating the illusion of heat, it does no permanent damage unless you eat a ton of chillies. And the release of endorphins while you are eating makes the rest of the food taste way more delicious than it is. This is why we are addicted to hot food.

The heat level of chillies is measured using the Scoville scale, which, I kid you not, involves extracting all the capsaicin in one chilli and diluting it with more and more sugar water till a panel of trained testers cannot detect any heat in the water. A typical green chilli requires dilution by 50 litres of sugar water for its heat to be imperceptible.

Like with most spices, chillies (both green and red) lose their flavour once they are powdered. They do not, however, lose heat because capsaicin is not volatile. If you want the flavour of the chillies, use them whole. If you only want heat, use the powder. If you are sensitive to heat, a common misconception is that it's the seeds that contribute all the heat. They don't. The seeds are removed because they taste bitter. It's the placenta, which connects the seeds to the flesh, that has most of the capsaicin. So, removing that will reduce the heat levels in your chillies. It's interesting to note that this misconception is yet another in the long list of 'wrong explanations with the right outcomes' that plague Indian home cooking. When you remove the seeds from a chilli, there is a good chance that you are likely using a knife to slice them away. The act of doing that will, in most cases, also slice away the whitish placenta to which the seeds are connected. So, it's natural to think that it's the seeds that contribute to the heat because the technique seems to work.

Chilli	Scoville Heat Units	Buckets (20L) of sugar water required to dilute the heat of 1 chilli
Capsicum	0	0
Kashmiri	2000	24
Byadagi	8000	95
Green Chilli	10,000	118
Gundu/Ramnad	20,000	236
Guntur	50,000	591
Kanthari/Bird's Eye	70,000	827
Bhut Jolokia	1,000,000	11820

Here is a cheat sheet for using heat in your dishes:

1. The right amount of heat intensifies other flavours.

2. Fat mutes heat, which is why idli gunpowder is paired with sesame oil or ghee.

3. Alcohol mutes heat, which is why bar snacks in India tend to be insanely spicy, because after a couple of large pegs, your TRPV1 receptors are not exactly in working condition.

4. Acid amplifies heat. Some years ago, my wife and I went on a trip to Sri Lanka. Being foodies, we asked our driver to take us to a place where the locals ate rice and fish curry. He did, and in case you haven't been to Sri Lanka, let me tell you that their recipes tend to start with the number of kilograms of chillies to be used, followed by other minor ingredients such as fish, tamarind, etc. Now I love hot food, but I can't say the same for my wife. The fish curry was so delicious that she somehow kept going, but at one point her brain said, 'Now hold on a minute, this individual seems to be continuing to switch on my TRPV1 receptors, and frankly, I'm tired of playing this annoying pain and pleasure game.' And so it decided to turn on the pain dial. As the restaurant staff saw my wife in great discomfort, they offered her some Coca-Cola, which, as I'm writing this book about food science a decade later, makes for a useful anecdote about not using acids to mute heat. Coca-Cola is highly acidic and will only makes things worse. After the Coca-Cola made things worse, the entire restaurant was invested in rescuing my wife and, finally, a serving of *vattalappam*, a rich coconut custard, which had enough fat to wash off the capsaicin, did the trick.

5. Heat alleviates richness or fattiness in dishes. When your dishes are too greasy, creamy or heavy, heat will reduce the perception of richness.

Origins of Flavour

As we discussed briefly earlier, the strongest flavours we use in the kitchen tend to come from the defence mechanisms of plants against microbes, insects and hungry herbivores. These are what we tend to generically term as *spices*. And because we cook our food, humans are unique in their ability to take what are largely nasty molecules to the rest of the animal kingdom and turn them into an explosion of flavours in our mouths using the application of heat and their combination with other ingredients.

For the sake of simplicity, because our primary interest here is arming ourselves with enough knowledge to make delicious food and not a PhD thesis in the taxonomy of flavouring ingredients, we will use a broad brush and categorize all strong flavouring ingredients into four buckets:

1. Dry spices: Clove, cardamom, black pepper, cumin, etc.

2. Fresh spices: Garlic, ginger or fresh turmeric.

3. Dried herbs: Fenugreek (kasuri methi) or dried mint.

4. Fresh herbs: Coriander leaves, curry leaves and mint.

You might wonder if I consider onions to be a fresh spice like garlic. I'd say it doesn't matter. Membership to this taxonomy should be based on whether it helps you be productive in the kitchen. For me, because onions tend to be used in relatively large quantities in most dishes, they are a common base-flavouring vegetable as opposed to an intensely flavoured spice.

This simple categorization will help you organize your spices—dried spices last the longest and can be stored in airtight containers for several months, while fresh herbs are the most perishable and even refrigeration will cause them to wilt and lose flavour in a few days.

But as a culture that uses more spices than any other people on the planet, it's useful to go one level deeper and understand where the flavour of spices comes from. Dr Stuart Farrimond, in *Science of Spice,* outlines twelve categories of flavour chemicals that we use in our cooking. The two major families of molecules behind almost all flavouring in our food are terpenes (the ones behind floral, woody and citrusy flavours) and phenols (the ones that impart stronger, unique and sometimes pungent flavours).

1. Sweet, warming phenols: Clove and fennel

2. Warming terpenes: Nutmeg and mace

3. Fragrant terpenes: Coriander

4. Sweet-and-sour acids: Amchoor

5. Fruity aldehydes: Sumac

6. Toasty pyrazines: When any spice is dry-roasted

7. Earthy terpenes: Cumin, nigella

8. Penetrating terpenes: Cardamom

9. Citrus terpenes: Lemongrass

10. Sulphurous and meaty: Garlic, black salt, mustard, asafoetida, curry leaves

11. Pungent: Chilli, black pepper, ginger

12. Complex flavour: Saffron, turmeric and fenugreek

Individual spices do not have just one of these flavour molecules. They tend to have several, with one or two dominating in terms of intensity. Sometimes two spices share some of these molecules, which makes them combine well with one another in a dish, a theme we shall explore later in this chapter.

A key thing to remember is that the flavours of spices work only in combination with the following seven basic elements:

1. Salt: Amplifies spice flavours

2. Sweet: Amplifies spice flavours

3. Bitter: Many spices by themselves have a bitter taste, but they also have a strong aroma that our noses detect

4. Sour: Balances spice flavours and makes multiple flavours stand out

5. Umami: Makes spice flavours linger in the mouth

6. Heat: Makes spice flavours pleasurable

7. Fat: Transports flavour. Most flavour molecules are not water-soluble. If not for fat, most flavours would simply be lost to the air.

Spice	Flavour Molecule	Flavours	Pairs with
Saffron	Picrocrocin Safranal Cineole Pinene	Musky, earthy, warm, bitter, Honey-like, floral, Penetrating Woody	Shahjeera, pepper coriander, cinnamon Ginger Garlic
Poppy	Hexanal, 2-Pentylfuran Limonene Pyrazines	Fruity, Green Grassy Citrus, herby Nutty, caramel, smoky	Chillies Coriander

Spice	Flavour Molecule	Flavours	Pairs with
Shahjeera	S-Carvone	Spicy, menthol	Star anise, cinnamon
	Limonene	Citrus, herby	Cardamom, pepper, ginger
	Sabinene	Citrusy, peppery, woody	Nutmeg, mace
Coriander	Linalool	Floral, citrusy, sweet	Cardamom, nutmeg, mace
	Limonene	Citrus, herby	Ginger, shajeera
	Pinene	Woody, spicy	Pepper
	Cymene	Fresh, citrus, woody	Cumin
Cumin	Cuminaldehyde	Earthy, herby	Cinnamon, cardamom
	Pinene	Woody, spicy	Pepper, black cardamom
	Cymene	Fresh, citrus, woody	Coriander, ajwain, nigella, star anise
Nigella	Nigellone	Mild, peppery, herby	Onion
	Limonene	Citrus, herby	Coriander, shahjeera
	Pinene	Woody, spicy	Pepper, cinnamon
	Cymene	Fresh, citrus, woody	Cumin, ajwain, nutmeg
Black Cardamom	Cineole	Eucalyptus, smoky	Cardamom
	Eugenol	Medicinal, woody, warming	Nutmeg, cinnamon
	Limonene	Citrys, herby	Coriander, shahjeera
Cardamom	Cineole	Herbal, penetrating	Black cardamom, nutmeg
	a-Fenchyl Acetate	Sweet, minty herbal	
	Limonene	Citrus, herby	Ginger
	Linalool	Floral, citrusy, sweet	Coriander, lemongrass
Lemongrass	Citral	Citrus, herby, zesty	Coriander, ginger
	Myrcene	Spicy, peppery	Pepper
	Linalool	Floral, woody, sweet	Cardamom

Spice	Flavour Molecule	Flavours	Pairs with
Turmeric	Cineole	Penetrating	Star anise, nutmeg
	Citral	Citrus, Herbal	Cardamom, coriander
	Zingiberene	Pungent, sharp, spicy	Ginger, pepper
Fenugreek	Sotolon	Bittersweet, caramel	Cinnamon
	Caryophyllene	Woody, spicy, bitter	Clove, Curry leaf, pepper
	Pyrazines	Nutty, roasted, smoky	Cumin, turmeric
Cinnamon	Cinnamaldehyde	Warm, spicy, pungent	Cumin
	Linalool	Floral, woody, spicy	Cardamom
	Eugenol	Medicinal, woody, warming	Clove
	Caryophyllene	Woody, spicy, bitter	Pepper
Clove	Eugenol	Medicinal, woody, warming	Cinnamon, fenugreek
	Caryophyllene	Woody, spicy, bitter	Pepper
Star Anise	Anethole	Sweet, liquorice, warming	Nutmeg, mace, cinnamon
	Cineole	Medicinal, penetrating	Black cardamom, ginger
	Phellandrene	Green, peppery, citrus	Pepper
Fennel	Anethole	Sweet, liquorice, warming	Nutmeg, mace, cinnamon
	Limonene	Medicinal, penetrating	Cardamom
	Pinene	Green, peppery, citrus	Pepper, cumin
Nutmeg	Myristicin	Woody, warm	Ginger, pepper
	Geraniol	Rosy, sweet	Curry Leaf
	Eugenol	Medicinal, woody, warming	Cardamom, clove
	Sabinene	Citrusy, peppery, woody	Garlic
Mace	Sabinene	Citrusy, peppery, woody	Garlic, pepper, curry leaf
	Terpineol	Floral, citrusy	Coriander
	Safrole	Sweet, warming	Star anise
	Eugenol	Medicinal, woody, warming	Clove

Spice	Flavour Molecule	Flavours	Pairs with
Asafoetida	Sulphides	Oniony, sulphurous	Garlic, mustard, nigella
	Ocimene, Acids	Sour, citrus, floral	Coriander
	Phellandrene	Peppery, Minty	Pepper, cumin
Curry leaf	l-Phenylethanethiol	Sulphurous, meaty, floral	Asafoetida, garlic, mustard
	Pinene	Woody, spicy	Nutmeg, pepper
	Linalool	Floral, woody, sweet	Coriander, lemongrass
	Cineole	Eucalyptus, herbal	Black cardamom, cardamom
Mustard	Isothiocyanates	Hot, peppery, penetrating	Chilli, garlic, ginger
	Pinene	Woody, herbal	Cumin
	Pyrazines	Nutty, roasted, sweet	Nigella
Pepper	Piperine	Hot, pungent, spicy	Chilli, mustard, ginger
	Pinene	Woody, spicy	Black cardamom, nutmeg
	Limonene	Citrus, herby	Cardamom, turmeric
	Myrcene	Piquant	Cinnamon
Ginger	Gingerol	Hot, pungent, spicy	Chili, pepper
	Linalool	Floral, sweet, herbal	Cinnamon, nutmeg
	Citral	Citrus, herbal	Lemongrass, coriander
	Cineole	Medicinal, penetrating	Cardamom
Ajwain	Vinyl Amyl Ketone	Earthy, creamy	Cinnamon
	Thymol	Cooling, penetrating	Nigella, nutmeg
	Myrcene	Peppery	Pepper
	Cymene	Woody	Coriander, Cardamom

Extracting Flavour

Now that you know where flavour comes from, and what kinds come from which spices, the next step is to understand how to extract flavour. How you choose to extract flavour has a significant bearing on the amount of flavour extracted. Depending on the dish and personal preferences, you might want a specific spice to impart a mild, medium or strong flavour. If every spice you use were to impart a strong flavour, your dish would be overwhelming.

The flavour of a spice in a dish depends on:

1. How you mechanically damage it: Cut, chop, smash, mince, grind, etc.

2. How you cook it: Dry-roast, oil-roast, boil in water, etc.

3. How long you cook it: In general, the longer you cook, the lesser the intensity of the flavour of spices, but this is a tricky concept. When you cook a gravy for a long time, the amount of water will reduce, which will increase the concentration of spices in your dish, thus making it taste more intense. So, it's important to understand this distinction. The intensity of a single spice's flavour will reduce with cooking, as more aroma molecules are lost to the air, while the dish in totality might taste more intense because it is becoming thicker.

4. What you pair it with, in terms of acid, fat, salt and sugar.

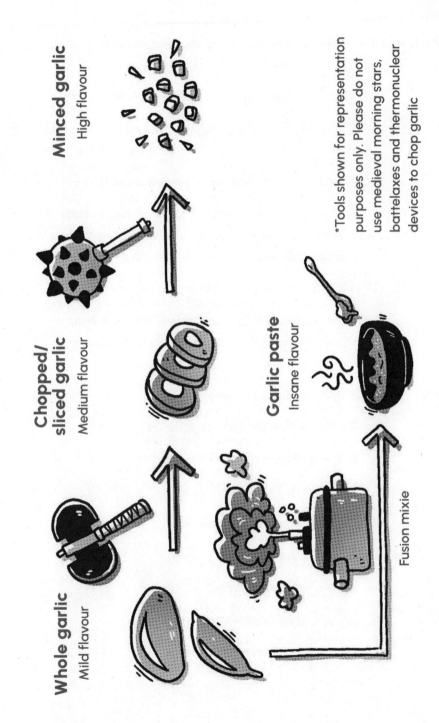

Minced garlic
High flavour

Chopped/sliced garlic
Medium flavour

Garlic paste
Insane flavour

Whole garlic
Mild flavour

Fusion mixie

*Tools shown for representation purposes only. Please do not use medieval morning stars, battelaxes and thermonuclear devices to chop garlic

And for each of these four categories of spices, these are the methods available to mechanically damage them to release flavour:

1. Dry spices (black pepper, cardamom, cinnamon, clove)

    a. Use whole, as we do with mustard and cumin, for a mild flavour

    b. Coarsely grind them for medium amount of flavour

    c. Finely powder them to extract the maximum amount of flavour. However, these have short shelf lives because powders leak volatile aroma molecules very quickly.

2. Dry herbs (fenugreek or dried mint)

    a. They are already flaky and powdery, so at best, crush them before use (like fenugreek) to help extract more flavours.

3. Fresh spices (garlic and ginger)

    a. Use large pieces for the mildest flavour

    b. Roughly chop for medium flavour

    c. Mince for medium high flavour

    d. Use a paste for maximum flavour

4. Fresh herbs (coriander or curry leaves)

    a. Use whole for mild flavour

    b. Roughly chop for medium flavour

    c. Mince for medium to high flavour

d. Use a paste for maximum flavour, but remember that the leaves tend to have enzymes that will start degrading flavour the moment any damage happens. So, use right away or use the trick described in Chapter 1, where you blanch it for 30 seconds and shock it in an ice bath before grinding into a paste.

Some exceptions include delicate spices like saffron, which are best used by soaking them in milk to extract their complex flavours into the milk fat and then using them at the end of a dish. If you want the pungent heat of mustard, it needs to be soaked in water for a few hours before being ground into a paste. Otherwise, whole mustard will simply provide a nutty texture contrast in your dish, nothing more. Powdered mustard will have some mild pungency, but not as much as soaked mustard.

When it comes to dry spices, one specific step tends to happen before the mechanical damage happens, and that's dry- or oil-roasting. The application of dry heat to the spices wakes them up and, for many spices, the amount of flavour you can subsequently extract from a dry-roasted spice is almost three times more than the unroasted spice. But be careful not to use high heat and burn the spice, or you will end up with acrid and bitter flavours. Roasting in oil, prior to mechanical damage, is also an option, and because flavour molecules tend to dissolve in fats, oil-roasting and then grinding/powdering them will extract the maximum flavour.

Dry herbs are best added late in the cooking process as they tend to be delicate. Too much cooking will largely dissipate their flavours. This is why kasuri methi is added once a dish is almost ready.

Fresh spices, such as ginger and garlic, can be used to impart a wide range of flavour intensity based on how they are cut and cooked. Roughly chopped garlic will have a milder flavour than the paste, but garlic cooked from the start of the dish will taste milder than garlic that is added towards the end. So, in the dishes that you do want a strong garlicky flavour, add it closer to the end, but only use roughly chopped or whole garlic cloves instead of the paste, so that you do not overwhelm the dish.

Fresh herbs are the most delicate of them all and, in general, best added right at the end of the cooking process, although curry leaves, which have a very intense flavour, do tend to survive long cooking and still impart their characteristic citrusy and meaty flavour to a dish. But the best cooks, when making dishes that use curry leaves at the start, will also add some fresh leaves at the end so that you get a hint of the fresh flavour as well.

While this might seem like a lot to process, it's straightforward when you consider some practical examples:

1. Raw whole spices + no heat: Little or no flavour.

2. Crushed spices + little or no heat: Medium flavour, like cardamom powder added at the end of payasam of kheer.

3. Crushed spices in water + heat: Mild flavour, like cardamom or ginger in tea.

4. Whole spices + oil + long cooking + medium heat: Moderate flavour, like using panch phoran or mustard/cumin at the start of a dish.

5. Roasted whole spices + crushing + oil + low heat: Maximum flavour.

6. Powdered spices + oil + high heat: Causes burning, so please avoid. Use low heat instead.

7. Oil-roasted whole spices + crushing: Don't cook for too long. Best added at the end of a dish.

8. Whole spices + oil + high heat: Mild flavour. This is the idea of the tadka (tempering). It's not to overwhelm the dish with last-minute flavour but to add a whiff of it by using high heat, which destroys most of the aroma molecules but leaves behind just enough.

To summarize, if you want to extract more flavour:

1. Roast dry spices.

2. Mince or grind fresh spices into a paste.

3. If using the spices whole, add them at the start of the cooking process. If using a powder, use it towards the end of the cooking process. You will see a ton of cooking videos on YouTube that add a ton of roasted cumin powder at the start of the dish. You are better off adding a much smaller amount of this spice towards the end and achieving a similar effect.

4. Use fresh or dry herbs only at the end of the cooking process.

5. Spices cooked in oil will taste more intense than spices boiled in water.

So, now you know how to decide how much flavour you want to extract out of your spices. As a general rule, minimalist dishes (like say, aloo jeera) will maximize the flavour of cumin without adding a ton of other spices to compete with it. And what is the difference between a good aloo jeera and an amaklamatic (from the Tamil word amaklam, which means amazing) aloo jeera? Flavour layering.

A good cook will pick the right method to extract flavour from every spice being used in a dish, so that the end product reflects what he/she wants to highlight. A great cook will take highlighting to the next level by layering different intensities of a spice's flavour in the same dish. Let's consider aloo jeera. A good cook will heat oil and add cumin, which will extract the whole spice's flavours into the oil, lending it a strong cumin flavour. Then he/she will add the potatoes, along with a pinch of roasted powdered cumin, which will provide a milder background flavour, since it's mostly cooking in the moisture of the potato and not in the hot oil. And, finally, a tadka with cumin will provide a visual and textural hit of the spice without necessarily having an overpowering flavour. A great cook will do all this and then drizzle some cumin and coriander infused oil before serving. Flavoured oils are typically top-notch chefs' undisclosed trump cards and will be discussed in Chapter 7.

Combining Flavours

When you are short of time and are trying to put together a quick dal and rice dinner, you go searching in your pantry for the spices to use. Quite likely, you will find some branded garam masala from the previous century that smells like sand. Also, it won't be uncommon to find a spare packet of instant noodles masala and throw that into your dal. In fact, your dish will taste delicious because instant noodle spice mixes tend to be extremely well-thought-out combinations, and that is what we will be discussing now. Now that you know where flavour comes from and how to extract the amount you want from a single spice, it's time to figure out how to combine them to create exciting and addictive flavours. To do that, let's analyse some common instant noodle spice sachets and see how they manage to transcend regional culinary preferences.

The combination of coriander seeds, red chillies, black pepper, fenugreek and cumin in the masala is a familiar blend for folks in south India. It's sambar or rasam powder, depending on the proportion of the spices. The garam masala in it is a familiar finishing spice mix to a wide swathe of people across India, particularly more so in the north, east

GARAM MASALA

Coriander Nutmeg
Pepper **Cumin**
Cardamom Clove
Black Cardamom
Cinnamon Mace

SAMBAR POWDER

Coriander Asafoetida
Pepper **Red chillies**
Chana Dal Turmeric
Mustard Toor Dal
Fenugreek

BIRYANI MASALA

Red chillies **Coriander** Mace
Pepper **Cumin** Bay leaf
Nutmeg Star Anise Fennel
Shahjeera Cinnamon **Clove**
Black Cardamom Turmeric

CHAAT MASALA

Coriander Garam masala
Amchoor **Red chillies**
Cumin Dried Mint
Kala Namak

RASAM POWDER

Coriander Pepper
Red chillies **Toor Dal**
Cumin Turmeric
Asafoetida **Chana Dal**

CHAI MASALA

Pepper Fennel
Clove Cardamom
Ginger powder Nutmeg
Cinnamon

*The font size corresponds roughly to the proportions of each spice

99

and west. Then there is amchoor, which adds sourness, something that is quite critical to creating a balance between all the other spices. Also, there is sugar that amplifies and enhances all the other flavours without perceptibly adding sweetness. And then there is the corn starch, as a thickening agent, so that your noodles does not come out thin and watery, apart from the three magic ingredients: onion, garlic and ginger powder. These are typically fresh spices, but when we dehydrate them and turn them into powders, they turn into addictive flavour bombs. In fact, the use of garlic and onion powder is what gives consumer snacks that intense, addictive taste. These sachets of instant noodles spice mixes are, in my opinion, one of the subcontinent's greatest, albeit underappreciated, spice combinations. It manages to taste like everything in general, and yet nothing in particular. Those used to a diet of sambar and rasam will detect familiar notes, while those used to eating biryani will detect those notes thanks to the garam masala and onion powder.

The lesson here is that while there is a science to blending spices, and we shall examine that in detail shortly, it's important to not forget familiarity and nostalgia. Nostalgia and memory play a strong role in flavour perception. After all, the olfactory cortex is within gossiping distance of the emotion and memory cortex. Familiarity and nostalgia are enabled by the stunning diversity of culinary traditions in India. If you start a dish with mustard oil and add mustard, fennel, nigella, fenugreek and cumin, it will bring Bengali cuisine to mind, no matter what you do after that. And if you start with coconut oil, and add curry leaves, garlic, mustard and cumin, it will evoke Kerala. Combinations of flavours that have been used for centuries in specific regions will almost always be the first place to look for inspiration.

ZA'ATAR (MIDDLE EAST)

Cumin

Salt **Sesame**

Sumac **Oregano**

QUATRE ÉRCES (FRANCE)

Nutmeg Clove

Pepper

Dried Ginger

BAHARAT (TURKEY)

Coriander Dried Mint

Cardamom **Cumin**

Nutmeg Clove

Cinnamon **Pepper**

ADOBO (MEXICO)

Pepper **Chilli powder**

Oregano Garlic Powder

Onion powder **Cumin**

FIVE SPICE (CHINA)

Sichuan pepper

Star Anise Cinnamon

Clove Fennel

RAS EL HANOUT (NORTH AFRICA)

Cinnamon **Cumin**

Dried Ginger Red chillies

Pepper Clove Coriander

*The font size corresponds roughly to the proportions of each spice

101

Now that you know what spice brings what flavour to the table, here are some principles to keep in mind when creating your own spice mixes. For starters, you need to gear up to become a spice ninja. Get yourself a spice grinder (or a coffee grinder that you can use for spices), and always buy whole spices. Powdered spices lose their flavour very quickly. And when you are in south India, your powdered spices will undergo the same experience as the batsmen who faced the West Indies pace quartet of the early 1980s (without helmets, mind you).

You are always better off grinding spice mixes fresh and having full control of whether you want to dry- or oil-roast individual ingredients to highlight or mute specific flavours. In addition to a spice grinder, get a mortar and pestle made of granite too, since it offers the most amount of abrasive firepower to crush spices. A mortar and pestle will extract more flavour from fresh spices, such as garlic and ginger, because the high-speed blade of a blender ends up heating and partially cooking the spice.

Let's recap the twelve categories of flavours in spices:

1. Sweet, warming phenols: Clove and fennel

2. Warming terpenes: Nutmeg and mace

3. Fragrant terpenes: Coriander

4. Sweet-and-sour acids: Amchoor

5. Fruity aldehydes: Sumac

6. Toasty pyrazines: When any spice is dry-roasted

7. Earthy terpenes: Cumin and nigella

8. Penetrating terpenes: Cardamom

9. Citrus terpenes: Lemongrass

10. Sulphurous and meaty: Garlic, black salt, mustard, asafoetida and curry leaves

11. Pungent: Chilli, black pepper and ginger

12. Complex flavour: Saffron, turmeric and fenugreek

You can also refer to the table of flavours we saw just a while ago.

1. Pick one or two flavour categories, the ones you want to be dominant in your mix.

2. Pick a spice from each category.

3. For each of those spices, pick one more spice (from any category) that shares at least one flavour molecule. For example, if you picked black pepper, your second spice should be black cardamom, which too has limonene. Black pepper and black cardamom will work well together as they will reinforce the citrus/herby notes of limonene.

Of course, nothing stops you from adding more spices to your mix, but remember that the more you add, the less your mix will stand out, as too many flavours will make your dish taste intense without necessarily being unique. Like the instant noodles masala.

3 Brown, Baby, Brown

Let onion atoms lurk within the bowl,
And, half-suspected, animate the whole.

—*Sydney Smith*

Ogres Have Layers

Anyone who likes eating, and I'm quite confident that this constitutes a fair chunk of the world's population, knows that the magic colour range for cooked food is the spectrum from golden to brown. From coffee to chocolate to freshly baked bread to fried chicken, the chemical process that imparts this colour to food is also the one that transports ingredients to a plane of deliciousness, something their un-browned versions only dream about. French physician and chemist Louis-Camille Maillard was the first to describe what exactly was happening to the proteins and starches in the cooking vessel. It's been about 100 years and we are still in the process of describing what happens when you cook food at temperatures above 110°C. It is precisely this spectacular diversity of chemical reactions that lends the seemingly infinite range of flavours to food all around the world. We shall begin exploring this magical browning reaction with the humble onion.

As far as literary metaphors go, comparing people to onions is rather common. Onions have layers, just like people, and so on. But I have to break it to you that it's not a very good metaphor. For starters, food science now tells us that the more flavourful, and correspondingly more tear-inducing, parts of the onion are the outer layers, not the inner ones. So, the next time you chop onions keep this in mind before callously discarding too many of the outer layers. But we are getting ahead of ourselves here. We are still standing in our kitchen, knife in hand, chopping board in place and some onions contemplating their eventual slaughter.

Let's first consider the onion from afar. Famous chef Julia Child once said, 'It is hard to imagine a civilization without onions; in one form or another, their flavour blends into almost everything in the meal, except the dessert.' But India being India, we have some vegetarian cuisines, particularly from the western part, that achieve spectacular results without onions or garlic. But in all likelihood, the recipes to a vast majority of the dishes you cook will start with the seemingly simple instruction: Sauté onions in oil. A few will be a little more specific and ask you to 'sauté onions in oil till they turn translucent'. But there is a century of food chemistry, starting with Louis-Camille Maillard, according to which how you choose to sauté your onions can radically transform the flavour profile of the dish.

This bulb-shaped root of the genus allium is, like the auspicious Ganesha squiggle that religious Indians make at the top of an empty sheet of paper, the starting point of a significant number of savoury dishes in every part of the world. We don't quite know where the onion originated, but we do know from the fact that traces of onion were found in the mummy of Pharaoh Rameses IV that it's been around for a really long time. Funnily, the onion traces were found in the pharaoh's eyes, so we don't know if that was perhaps a bizarre case of revenge by a scorned mummification intern, or if there was a deep-seated conspiracy involving the switching of the real pharaoh's body with an unfortunate kitchen assistant whose thankless job was to chop onions all day.

Onion-eyed pharaohs aside, the ancient Egyptians considered the multi-layered spherical bulb-shaped vegetable to be a symbol of the universe. The Latin word 'unus', which means 'one' is the likely origin of the word 'onion'. In addition to the kitchen, the onion also played a role in folk medicine around the world, with purported curative properties for ailments such as colds and animal bites. I say purported because there is currently no peer-reviewed scientific evidence for any of its medicinal properties.

Onions are about 89 per cent water, 9 per cent carbohydrates and 1–2 per cent protein. They also contain very little essential nutrients. Essential nutrients are biochemical components that your body cannot synthesize either enough by itself, or not at all, and need to be eaten as part of one's diet, like some amino acids and most vitamins. Onions also contribute a very small number of calories per serving, which explains why they are such a widely used way to add flavour to any dish.

As a small aside, the software engineer in me has to tell you about Larry Wall's annual 'State of the Onion' speech. One of the most popular programming languages in the early days of the Internet was PERL, and as it was with most technology back then, PERL was a complex mess that kept adding more and more features as the appetite for building web applications developed over time. Its open source nature meant that developers kept adding layers to the base of the language and, thus, the annual 'State of the Onion' speech to all PERL enthusiasts, by its original developer and patron saint of the early Internet, Larry Wall, came to be.

So, now we are back in our kitchens, staring at our onions, and there are three of them. There's a yellowish-brown one, and it is the absolute best-tasting and most widely used variety in the world. Unfortunately, it is not very commonly available in India. Then there is a purple variant, which is great for use in salads but is not as flavourful as the yellow one. It is truly a testament to the sophistication of various cuisines in India that we make do with what is ultimately a substandard variety of onion and coax the most amazing flavours out of it. But this is a rather recurrent theme in Indian home cooking. We overcook vegetables, cook meat till it's dryer than the surface of the moon, and we like our eggs boiled harder than Pantera's music. However, we compensate for all of that with a flavour-bombing strategy that has no parallel anywhere else in the world. We tend to use a large number of contrasting flavouring ingredients, and rarely just onion or garlic. In contrast, consider the French onion soup, which relies entirely on two primary ingredients: lots of onions and meat stock. We rarely realize that most non-Indians look at recipes for Indian food and are overwhelmed by the ingredient list. And finally, there is a white variety of onion, used more commonly in Mexican and Middle-Eastern cuisine, which is distinctly sweeter in taste.

If you are from south India, you might wonder why sambar onions are missing in this discussion. The tiny miniature-looking things technically aren't onions. They are shallots, which have a more complex yet sweeter taste profile. South Indian and South East Asian cooking tends to prefer shallots over onions, as they pair excellently with hot, red chillies and curry leaves. If you can't find shallots, you can, mostly, replace them with onions, but a sambar or Penang curry made with onions instead of shallots does not quite taste the same. In addition to the shallots, there are also leeks, chives and spring onions, but they are less common in day-to-day use in this part of the world, and are, for the most part, oniony in varying degrees.

Now that you have determined which kind of onion (or shallot) is required for your dish, it's good to understand the elements of its internal structure that impact flavour. Onions have particularly large cells, which run along the bulb from the root, which is why they are commonly used in high-school science education, as you can see the cells under very low magnification. What this means is that how you choose to cut your onions will significantly alter their final flavour.

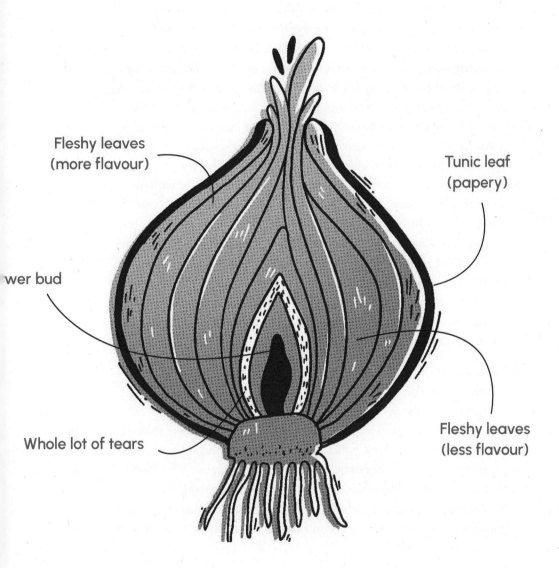

Fleshy leaves
(more flavour)

Tunic leaf
(papery)

wer bud

Fleshy leaves
(less flavour)

Whole lot of tears

When you cut an onion across the bulb, and not along it, you break the cells, which release, after a chain of reactions that takes about thirty seconds after mechanical damage to the cells, an aerosol of syn-Propanethial-S-oxide, a sulphur-based chemical that immediately causes the lachrymose (tear) glands to fight back by generating tears to flush the irritant chemical out. The onion, too, does this as a defence mechanism, because it has a vested interest in preventing microbes, insects and grazing animals from chomping on it. In fact, the entire allium family of plants has evolutionarily figured out that generating volatile, sulphur-based molecules is an excellent deterrent because when you want to make molecules that smell nasty and taste funny to animals, without necessarily being poisonous, sulphur is your friend. Garlic does this by converting a substance called alliin into allicin the moment Nandini, the cow, bites into a garlic bulb. Allicin tastes terrible to most mammals, but not to humans. We, for some reason, love these flavours, but only after we cook them. Nandini, the cow, is unfortunately missing out on this party. Incidentally, these sulphur-based defence mechanisms also damage the red blood cells of dogs and cats, so never feed your pets onions or garlic.

An interesting thought experiment: While the evolutionary defence mechanisms in the allium family exist to prevent animals from eating the plant, humans still bear with the daily ocular torture and post-meal olfactory assault to eat it on such a large scale. And, unlike animals, humans invented cooking, which neutralizes the reaction that causes us to tear up or causes dishes to be overwhelmingly pungent. One might then think that perhaps the defence mechanism ultimately failed, but consider this: Onion and garlic seem to have convinced human beings to grow them on a massive scale around the world, so who actually won this evolutionary battle?

But back to onions for now. Onion cells, when broken, release enzymes called alliinases, which in turn break down alliin to ultimately produce many flavour molecules that you uniquely identify with the taste of onions. It turns out that one of these many reactions produces the volatile chemical (syn-Propanethial-S-oxide) that rises into the air from the chopping board and assaults your eyes. Now, to this bit of chemistry knowledge, you can apply some simple engineering physics and reduce your chances of tearing up:

1. You could chop onions under running water. This is wasting of water though.

2. You could use a sharp knife. A dull knife will damage more cells than necessary.

3. You could keep a small USB fan near the chopping board to blow away the syn-Propanethial-S-oxide before it hits your eyes.

4. You could refrigerate the onions for a little bit before cutting them. Don't store onions in the fridge for long though because they need well-ventilated spaces, and fridges are not exactly ventilated. The science trick here is that if you remember your high-school biology, most cellular reactions require at least body temperature (37°C, or 98.6°F), so briefly chilling the onion before cutting it will significantly reduce the rate of that enzyme reaction. Temperature is, if you recall, a measure of how much energy molecules in a substance have. At colder temperatures, they are mostly chilling out and relaxing, and are not too keen on reacting with each other. As the temperature rises, molecules start jumping about and increase their chances of running into each other, exchanging electrons in the process.

Why Do Onions Make Us Cry

The chef cuts the onion.

When the onion is cut it releases its cells.

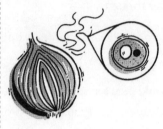

These cells burst and release Amino Acid Sulfoxide and an enzyme. They mix to form a gas at room temperature.

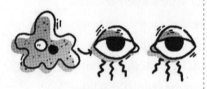

When the gas enters our eyes it creates a mild sulphuric acid that gives us a burning sensation.

A gland at the side produces a liquid to cut down on the burning.

When our eyes can't hold any more liquid they fall out making us cry.

Sulfoxide isn't an amino acid

When you cut an onion, you break open cell after cell, releasing their liquidy contents. Previously separated enzymes meet the sulfur-rich amino acids and form unstable sulfenic acids, which then rearrange into a chemical known as syn-Propanethial-S-oxide. This sneaky chemical is volatile, meaning it easily vaporizes, and causes a burning sensation when it floats up from the cutting board and comes in contact with your eyeballs. In an attempt to keep your precious eyes from being damaged, your brain quickly triggers a tear response

The primary flavour profile of an onion is savoury and pungent. In fact, the molecule that causes you to tear up is also the one that, when heated, is converted to another molecule called MMP (3-mercapto-2-methylpentan-1-ol, if you want to show off to your friends). MMP lends the meaty, savoury and luscious taste to any gravy that features onions. Now you know why so many gravies start with onions in this part of the world. The more mechanical damage you do to the onion, the more MMP is released, and thus more intense the flavour in your dish. It is also fairly water-soluble, so adding a bit of water after sautéing the onions will get you a stronger flavour, as it will prevent any further loss of aroma to the air.

Raw onions are also mildly acidic. The pyruvate scale measures the pungency of onion and garlic on a scale of 10. If you are using an onion for a salad and will be eating it raw, just soak it in water for a bit to remove most of the pungency. If you want your salad onions to remain crunchy, don't use warm or hot water, as the heat will break down cell walls and turn the onions limp.

The flavour of an onion in your dish depends significantly on how you choose to chop it in the first place, essentially how much cellular damage you do:

1. Slicing pole to pole: Least pungency

2. Slicing across the bulb: More pungency

3. Fine mince: Maximum pungency

As mentioned earlier, the inner layers of an onion are less pungent and flavourful than the outer layers. So, think twice before discarding too many of the outer layers. Incidentally, the enzyme reactions we just spoke about continue well after you cut an onion. They don't just stop because, at room temperature, the enzymes are still active. This is why it's not a good idea to cut onions in advance. They will lose flavour and also develop a slightly bitter taste, as the enzyme reactions continue unabated. Cut them as close to the cooking time as possible. If this is not practical, consider refrigerating after you cut them.

Now that we have chopped the onions in a manner that suits the flavour profile and texture of our dish, we must consider how to cook it. The vast majority of Indian recipes will begin with the instruction: Sauté onions in oil/ghee/butter. So, let's understand what happens when you add onions to hot fat. A quick recap from Chapter 1 will help here. Always heat the pan before you add the oil, and heat the oil before you add the onions. In general, you don't want to slow-heat oil because it will only oxidize more. You also don't want to add your onions before the oil is hot because adding anything to oil lowers its temperature. Another science tip to remember: You want to heat the oil to something way less than 177°C, which is 'frying' temperature. We are not frying onions at this point, although that is a perfectly valid thing to do if you are making biryani. And you don't need a thermometer to get this right either. If you hear a mild sizzle, your oil is ready. A loud sizzle indicates close to frying temperatures, while no sizzle indicates that the oil is not hot enough.

So, once you've added the onions to hot oil, and you hear the perfect sizzle, the clock starts ticking.

Stage 1

Sweating: The water inside the onion's cells starts to heat up, and since the oil is well above 100°C, it will start to turn into water vapour. This removal of water from the cells will cause the onion to start wilting slightly. It will, however, still have a bit of crunch and be opaque. If you add a pinch of salt at this stage, it will dehydrate the onion even faster.

Stage 2

Cook till translucent: At this point, the onion has lost most of its water, is fully wilted and has no crunch.

Stage 3

Light browning: As you keep cooking further, the onion will start to brown. This is the Maillard reaction. The amino acids in the onion (the 2 per cent protein, remember) will react with sugars (the 9 per cent carbohydrates) to produce several delicious brown compounds that we shall analyse in detail subsequently.

An interesting fact: The amino acids in your own cells react with sugars over time, in a very slow version of the Maillard reaction, to render proteins in your tissues dysfunctional, which, scientists say, is a component of human ageing! Imagine the irony in the fact that the same reaction that makes your onions delicious is also the reaction that slowly kills you. This is also why doctors keep telling you to keep your blood sugar levels low.

Stage 4

Golden brown: This is, and I quote singer–songwriter Kenny Loggins, the danger zone. The Maillard reaction is now in full tilt, and unless you have a good-quality pan and even heating, some of your onions are probably already burning, producing more bitter tasting by-products. At this stage, you should be babysitting the onions and constantly stirring them to prevent burning.

Stage 5

Caramelization: If you have the patience to go low and slow, you can take things beyond Stage 5 and produce one of the greatest onion products of all time—caramelized onions. There's just one problem. The name. It's a bit of a misnomer. As we shall learn later in this chapter, caramelization is an entirely different chemical reaction that happens to sugars at extremely high temperatures, where they break down into a family of caramelly and nutty-tasting molecules. Caramelized onions, on the other hand, are just onions Maillardized to their fullest possible potential, with the cell walls fully breaking down and the texture jammy. That said, if you keep heating onions, the sugars will eventually caramelize too, at which point the flavour profile will be more complex, caramelly and less oniony. This is when the onions look really dark brown. For most novice cooks, I'd stop short of too much caramelization.

Caramelized onions are a fantastic condiment used regularly in Western cooking, as a spread on bread, or on top of burgers. They are also the key to making a delicious French onion soup. A disappointingly flat-tasting French onion soup is one where the onions are not caramelized enough.

We can apply this new-found knowledge to Indian cooking too. Adding roasted cumin powder, chilli powder and chaat masala during the caramelization stage will give you a unique alternative to tamarind chutney. You can even add amchoor to add an element of sourness to what is a complex, sweet flavour. You can alternatively take caramelized onion (or onion jam, as it's sometimes called) and blend it with red chillies, roasted peanuts and grated coconut to make a fantastic chutney to go with dosas and idlis.

That said, it does take a fairly long time for onions to get to Stage 5, typically up to 45–60 minutes if you are cooking about 500 g. Worse, it takes near constant attention and stirring to keep them from burning.

So, while the food chemist will say 'good things come from letting time, heat and Dr Maillard do their work', the software engineer in me wants shortcuts, parallel process-ability and automation.

It turns out that you can pressure-caramelize onions! All you have to do is add butter, salt and onions to a pressure cooker and cook it for 20 minutes at peak pressure. The onions have enough water in them, so you don't need to add more water. If you are nervous, go ahead and add a teaspoon just to soothe your anxiety from trying water-free pressure-cooking. The butter will, in any case, prevent the onions from getting scorched. You won't necessarily save too much time with this technique, but you won't have to babysit the onions. The catch is that pressure-caramelized onions don't taste as good as the 60-minute open-pan version because Dr Maillard likes time and temperature. The peak temperature in a pressure cooker is 121°C, and some of the best-tasting steps of the Maillard reaction happen around 150°C.

So, if you do not want to use a pressure cooker, there's a magical chemistry trick you can use to accelerate the Maillard reaction, particularly when cooking onions. The magic ingredient is sodium bicarbonate, also known as baking soda. Unfortunately, baking soda, like MSG, has an image problem thanks to wholesale misinformation over the years. Here is the unvarnished scientific truth: using baking soda or MSG in small quantities has no proven negative effects on health. That said, baking soda, which is mildly alkaline, tastes terrible. Also, unused sodium bicarbonate reacts with the hydrochloric acid produced in your stomach to generate what in polite company tends to be called bidirectional body wind. How is it magical then, you might wonder? If you recall, onion cells, and all plant cells actually, are made of a polysaccharide (multiple sugar molecules attached together) called pectin, which provides structural integrity to the cell. So, when a vegetable goes limp in the cooking pan, it means that most of the cells have had a pectin breakdown. But pectin is a pretty hardy molecule that doesn't go down without a fight. Here is where sodium bicarbonate comes in. Like Vidkun Quisling during World War II, it accelerates the breakdown of pectin in plant cell walls.

You can do this experiment at home. Finely chop some onions and add it to some hot oil in a pan. Now assume you are Great Britain (circa 1947) and partition the onions into two portions. In one half, add a tiny pinch of baking soda. Stir things around to prevent the onions from burning, while keeping the line of control intact. In a few minutes, you will see the onions with the baking soda turn a delicious golden brown, while the other side remains mostly translucent. Baking soda accelerates the breakdown of pectin, which in turn releases the proteins and sugars inside the onion. These, when heated up, undergo the Maillard reaction. As long as the pectin holds, it will guard the innards of as many onion cells as possible with great dedication. This, unfortunately for us, results in under-flavoured onions in our dish. But baking soda has another trick up its sleeve. Maillard reactions happen faster in alkaline environments, and sodium bicarbonate raises the pH level of anything it's added to, all of which has the effect of a single precision-guided stone missile that de-fruits two mangoes from a tree. Not only does it break down pectin, it also hastens the browning of onions. You can also use soda to great effect when trying to save the planet by reducing the use of energy to cook lentils, particularly chickpeas and black urad dal. A pinch of baking soda added to the lentils in the pressure cooker will cook them in half the time it would otherwise take. It's the same chemistry.

Baking soda breaks down the pectin in the tough exterior of the lentils and allows water to enter the seeds, causing them to expand in size as the starches gelatinize, making the seeds edible. In fact, it is the occasional overuse of baking soda by the low-cost restaurant industry in India that has given soda a bad name. But what is a blunt tool to save on LPG for the restaurant industry is a precise Swiss knife in the home kitchen. So, ignore all those WhatsApp forwards and always keep a big box of sodium bicarbonate in your kitchen. In addition to its Maillard-acceleration and pectin-destruction skills, it is also a fantastic cleaning agent. The Internet is filled with videos of people demonstrating the magical combination of soda and vinegar for cleaning kitchen surfaces, but here is what chemistry tells us. Both soda (base) and vinegar (acid) are good cleansers individually. When you mix them, you get a frothy, bubbly reaction that produces good old carbon dioxide and sodium acetate, which is the salt used by the big brands to lend that lovely salt-n-vinegary flavour to potato chips. Sodium acetate is abrasive, which means that when rubbed against surfaces, it will dislodge dirt and grease stuck to them. So, you can simply use vinegar by itself, or baking soda and water, or a combination, whose dramatic frothing effect provides the illusion of cleaning although it's just carbon dioxide being produced.

After that mild detour, let's get back to our onions. It turns out, to summarize, that onions don't just have physical layers, they also have layers of flavour that you can unlock depending on how you chop them and cook them. In general, the idea is to think about what flavour you want.

If you cook your onions till they are translucent, they will impart a mild flavour to your dish. I'd use this for milder, creamier curries, like kormas, etc. Mildly browned onions will impart a complex sweet and savoury flavour, suitable for tikka masala-type strong-flavoured gravies.

If you go all the way to golden brown, the onions become suitable for bhuna (roasted), rogan josh or theeyal-type dishes. If you go all the way and caramelize them, you might as well just go ahead and smear it on a piece of toast and wonder what you've been missing out on all these years. Or you could get creative and use it in Gujarati-style preparations that call for a distinct sweeter note, in addition to the spicy ones. Instead of just using jaggery or sugar, use caramelized onions to get a fantastically more complex-tasting Gujarati dal.

Science of Garlic

Many of the food science ideas we have discussed about onions apply to every other ingredient. Garlic, for instance, is quite similar. It comes from the same family of plants and has similar sulphur-based volatile flavour molecules that are released when its cells break down. The big difference is that these volatile compounds are not eye irritants, just spectacularly smelly. One of the compounds produced when you cut into garlic is allyl mercaptan, a close cousin of ethyl mercaptan, a chemical so smelly that our noses can detect even one molecule in an entire room. This property has a very useful application. When added to LPG, which is dangerously odourless, it helps us detect gas leaks in our kitchens. Single molecules of ethyl mercaptan have saved millions of lives over the years. While most of the smelly compounds garlic produces are quickly broken down by our digestive systems, one notoriously resilient molecule named allyl methyl sulphide is very hard to break down. In fact, much of it passes unchanged through your digestive tract and enters your bloodstream, from where it is typically excreted via your urinary tract and sweat. So, remember, your entire body will have a lingering, garlicky aroma for up to twenty-four hours when you eat a ton of garlic. Fortunately, we don't need a ton of garlic to add flavour to dishes, because a little goes a long way. As with onions, how you choose to cut it and cook it will determine the intensity of flavour.

The less cellular damage you do to the garlic, the less garlicky its flavour will be. So, whole garlic cloves will lend a milder flavour than roughly chopped garlic, while minced garlic will be the strongest of them all. But cooking garlic is much trickier than onion. It has much less water content

(59 per cent) than onion (89 per cent), making it way more susceptible to burning. And if you are wondering what water content has to do with this, it's time for a quick recap from Chapter 1. Every substance has a specific heat capacity, which is a measure of how much energy you need to increase the temperature of a fixed amount of that substance by 1°C. Metals, as you might imagine, have lower specific heat than water. This is why your cooking vessels are made of metals in the first place. They heat up really fast, while the water or milk you are trying to boil seems to take forever.

But let's be clear. Higher water content does not automatically imply slower cooking time. In fact, pumpkins, which are mostly water, cook in much less time than potatoes, which have less water, so it's important to understand the distinction between cooking and burning. Cooking is the strategic application of heat to transform an ingredient into a narrow range of acceptable flavours and, more importantly, textures. Burning is the brute application of heat with the malicious intent of turning your ingredient into elemental carbon.

What happens when you cook garlic is that its relatively smaller size and lower water content increase its chances of burning sooner rather than later, which is why garlic needs a lot more babysitting than onion. That said, the higher percentage of sugars in garlic (30 per cent) means that it will brown more effectively and quickly as the sugars undergo the Maillard party with its proteins (6 per cent). But browning garlic is very tricky business because the line between getting good flavours from the Maillard reaction and nasty, bitter flavours because of burning is very slim. This is one reason why it's generally safe to avoid browning garlic when it's part of a more complex dish, because the chances that it will add bitter flavours to your dish is very high, unless your nose is finer than the one on a French sommelier. A simple rule of thumb: When you add garlic to a pan, always keep the heat at medium–low. And if you observe your grandmother, she will likely add the garlic only after the onion. The reason for this is that onion, which has more water, releases it into the pan, which keeps the garlic from burning. So, if your recipe calls for garlic without onion, be extra careful.

Now that you are armed with all this knowledge, you can experiment with making browned garlic chips, which make for the most delicious garnish. Deep-frying chopped garlic for just the right amount of time to Maillardize the outer layer also makes for a fantastic tempering that can be added to dishes like lasooni dal, or garlic rasam.

But garlic is a polarizing flavour in the Indian subcontinent. There are entire communities in India that do not eat garlic at all. Whether this has to do with the fact that one of the primary flavours of garlic (and even onions, to an extent), particularly after it has undergone the Maillard reaction, is a savoury, meaty flavour similar to what you get when you grill meat, we do not know, but Ayurveda has slotted this delicious vegetable in the *tamasik* (qualities that obstruct us from understanding happiness and wisdom, and encourage inertia, ignorance, etc.) category. Even those who eat garlic avoid it during religious festivals. As we learnt in Chapter 2, there is an entire category of spices that uses sulphur-based molecules to impart a savoury and meaty flavour to dishes. Curry leaf is another example. In fact, its name comes from the Tamil word for a meat-based dish called kari.

Coming back to garlic, I want to introduce you to this scientific idea of a calibration exercise. Chop garlic into various sizes (whole, rough, minced, paste) and sauté it for varying amounts of times at medium–low heat. Then get your family members to taste each one and rank their preferences. Tabulate these results and decide what the perfect garlic threshold is for your family. The complex and rich flavour profile of perfectly cooked garlic in a dish cannot be replicated by any other ingredient. This exercise will achieve optimal garlic harmony in your house. Consequently, everyone in your house might be leaking allyl methyl sulphide from their skins for the next twenty-four hours, but it is *for science.*

Beyond the optimal cooking of garlic, you might wonder if it's possible to get it to Stage 5—extreme Maillardization. It is! But you can't do it on the stove because the garlic will most certainly burn well before we get to that stage. To caramelize garlic, you will need a convection oven (or an OTG). Here, it's time for a quick recap from Chapter 1 again. We learnt that water needs more heat than other substances to warm up by 1°C (specific heat capacity). But there is another property called thermal conductivity, which is a measure of how well a material can transfer heat to something else. Water is an excellent conductor of heat compared to air. This is why a room at 30°C feels stifling, while water at 35°C feels pretty comfortable. The water will conduct heat away from your body much better than the air in the room will. When it comes to cooking, this is why things cook faster in a watery gravy than they do in the oven, where air is the medium of cooking. But we can utilize the inferior conductivity of air to our benefit when it comes to caramelizing garlic.

A whole bulb of garlic (mind you, cloves of garlic will likely burn), drizzled with some oil, wrapped in aluminium foil in a 175°C oven, which is the temperature at the end of the Maillard reaction chain, for 30 minutes will yield spectacularly caramelized garlic which, when added to butter, makes for the most satisfying garlic butter you will ever taste. The papery outer layers of the garlic bulb and the individual cloves will mostly be burnt at the end of 30 minutes. In fact, it serves as an additional layer of protection to prevent the flavourful innards from burning.

Caramelized garlic can also be used in chutneys to add a more complex and sweet flavour, compared to just raw or pan-browned garlic. Try this: Blend caramelized garlic, cashew nuts, salt and green chillies to get the most astonishing chutney with the complex, savoury and sweet taste of garlic, the creamy and nutty texture of cashew nuts, the herby hot freshness of green chillies, and salt to bring it all together for a flavour explosion in your mouth.

By the way, if you want to deal with garlic breath using some knowledge of chemistry, eat parsley, apples or pears, which contain our good old friend polyphenol oxidase. Yes, it's the same guy that makes our green vegetables lose colour by stealing magnesium from the chlorophyll molecule. Polyphenol oxidase, when exposed to oxygen, reduces the volatile compounds in garlic and, therefore, garlic breath. Other things you could consume are lettuce, peppermint, basil, and mushroom, as these are also effective in removing the methyl mercaptan and allyl mercaptan, the nasty-smelling compounds in garlic.

Cabbage and Potatoes

A vegetable that I (and several millions more) used to abhor as a kid is cabbage. It turns out that, more often than not, cabbage is a victim of overcooking. When you boil cabbage, as opposed to Maillard-browning it, it releases hydrogen sulphide, a gas we are familiar with from our high-school chemistry labs. It smells like rotten eggs, and I might be going out on a limb here, but I think most people don't fancy that smell in their food. But urban Indian kitchens tend to be wary of undercooking anything for obvious historical reasons—we don't know what went into growing, storing and transporting the ingredients, and raw ingredients have historically been recipes for bacterial invasions in our bodies. So, there is a natural tendency to err on the side of overcooking and compensate by adding a ton of spices for flavour, along with tempering with roasted lentils to compensate for the loss of crunch.

But cabbage is not too different from the onion, in the sense that it is mostly watery but packs a ton of flavour molecules, which can be unlocked with the slow application of heat. If you are patient, you can, in about 30 minutes, brown and caramelize cabbage into something way more delicious than steamed cabbage. Here's a quick recipe for caramelized cabbage sabzi: Caramelize the cabbage separately while you make an onion, ginger, garlic, chillies and spices base. Add the browned cabbage to this and quickly mix it before turning the heat off. Temper with mustard/chillies. Caramelized cabbage has more flavour than onion because it has more glutamates, which lend a umami (savoury) flavour to the dish. This idea will be explored in Chapter 5.

But no discussion on browning can be complete without the king of vegetables, that underground wonder tuber whose discovery by the Europeans in the Andean highlands helped them turbocharge colonialism. Yes, the humble potato, it turns out, was the most efficient way for the colonizers to carry carbohydrates on ships, because it doesn't spoil quickly and provides reasonably balanced nutrition. So, once the Europeans figured out how to grow this tuber in their part of the world, it meant dark times for literally every part of the world that was eventually colonized. But Indian cuisine wouldn't have included potatoes had it not been for the Portuguese colonizers. In fact, the Marathi word for potato is literally the Portuguese word: '*batata*'. Uncomfortable history aside, you can use your expert knowledge of the Maillard reaction to make the perfect, golden-roasted potato by simply adding a pinch of baking soda to the vessel in which you are boiling the chopped potato. Baking soda will break down the pectin on the outer layers and create ridges and grooves that, when encountering hot oil, will turn into the most scrumptious and crisp exterior, even as the insides remain perfectly soft and cooked. Make sure you wash the potatoes once they are parboiled because unused baking soda does not taste very nice.

There is a reason why it's so hard to get potatoes to be the perfect golden brown in India. Blame it on the high levels of moisture in most Indian varieties. Most commonly available potato varieties in India, particularly in the south, tend to be the low-starch, high-moisture variants that are good for gravies, but not for French fries or chips. So, when you try to Maillardize these chaps, the liquid water in them says, 'Excuse me, as long as I am around, I am not going to let the temperature climb over 100°C.' And till all the surface water becomes vapour, you can't get the Maillard reaction going! The food industry grows a special high-starch, low-moisture variant called chips potatoes. These are not available in the retail market for complex agricultural policy reasons we will avoid here.

Maillard Reaction

We've been talking about the Maillard reaction for a fair bit now. It's time to understand what is happening here. In the 1920s, French physician and chemist Louis-Camille Maillard described a series of reactions that were central to turning proteins and carbohydrates into delicious, easier-to-digest compounds. Our focus shall primarily be on the 'delicious' part of the story, because the easier-to-digest part is for books about nutrition. Anything deliciously brown, from seared meat to ground coffee to chocolate to caramelized onions, garlic or cabbage, has undergone the Maillard reaction.

For the purposes of this book, we will dispense with a lot of detailed chemistry. Instead, we will look at what happens to the ingredients in your cooking vessel as their temperature hits 110°C, which is when the Maillard reaction starts. This will start to happen only when all the water on the surface of the ingredients has boiled away. Remember, as we learnt in Chapter 1, this reaction requires a dry cooking process.

It's a complex chain of reactions where the output of one interacts with the output of another to produce entirely new molecules, which then transform into other molecules. What is important for us to remember is that between 110°C and 170°C, the conversion of sugars in carbohydrates and the amino acids in proteins into an unstable molecule called N-Glycosylamine happens. This unstable molecule rearranges itself into super-flavourful molecules like:

1. Furans: These lend a deep, slightly burnt and caramel-like flavour.

2. Furanones: These lend a subtle sweet taste we associate with browned foods, such as the crust of a bread.

3. Oxazoles and pyrroles: These lend nutty flavours.

4. Thiophenes: These add a meaty, savoury and roasted flavour.

5. Pyrazines: These add a toasty flavour, like when browning sesame seeds.

6. Melanoidins: These impart the quintessential brown colour to food. This family of molecules is the reason why coffee and chocolate are brown.

Every time you take an ingredient that has both sugars and proteins, and let it heat up to 110°C, you unlock a combination of the flavours listed above. Of course, this varies based on your starting material. When you brown any ingredient, you must also leave it unmoved on the pan for a little bit because contact with hot metal is what gets the temperature high enough. In an oven, browning takes time because air is a terrible conductor of heat, but you sure can brown more evenly.

roasty and toasty
Pyrazines

Herby, nutty
Oxazoles

nutty
Pyrroles

fresh-baked
bread, caramelly
Furanones

savoury, caramelly
Furans

roasted, meaty
Thiophenes

Amino
acids

Heat
110 – 154°C

Sugars

So, here's the biggest food science tip for the Indian kitchen, albeit one that restaurants use all the time and good cooks just seem to know. Consider browning as many ingredients as possible *before* adding them to gravies, if you truly want the most delicious flavours. A simple example: Roasting pumpkins in an oven, or on a grill, before using them to make pumpkin soup makes for a significantly more delicious soup than simply boiling them in a broth. You can use this principle in any dish you wish to make, for example, cauliflower gravy. Lightly steaming the cauliflower and then browning it in butter before adding it to a tomato-based gravy will yield a distinctly superior-tasting final product than simply dropping the cauliflower into the gravy. Chapter 7 will explore the art of prepping ingredients in great detail.

But do remember this, not all products of the Maillard reaction are good for you. At high-enough temperatures, one of the by-products of this reaction is acrylamide, which is bitter and, more importantly, carcinogenic. So, when people warn you about not eating burnt food, this is exactly what they are talking about.

Caramelization

A slightly confusing thing about this chapter is the liberal use of the word caramelization to refer to things that didn't actually undergo much caramelization. The only reason for this is that while we must pay attention to scientists in most situations, we shouldn't let them '*well, actually*' us into changing terms people understand. Clarity of communication and utility always trump technical correctness. Most cooks know what caramelized onions mean, so those correcting them with a 'well, actually, it's Maillardized onions because caramelization is an entirely different chemical reaction' deserve to be punched in the face.

What is useful for us though is to understand what the actual caramelization reaction really is. Unlike the Maillard reaction, which happens between amino acids and sugars, caramelization happens to sugars when they are heated to a very high temperature. When you heat onions at low heat for a really long time, you will eventually get to the point where both the Maillard reaction and caramelization of sugar happen, but that is dangerously close to burning territory, so be very careful. The more you actually caramelize the sugars in the onions, the sweeter they will taste.

What you can safely, fully caramelize is actual sugar. When you heat sugar slowly (add a little bit of water to ensure even heating), it will first melt and then start turning brown around 160°C. At this point, a plethora of aromatic molecules are produced that result in the familiar caramel flavour. Different sugars caramelize at different temperatures. Fructose, found in fruits, tends to caramelize at 110°C, which is why caramelized bananas are popular in desserts. Sucrose, which is your common white sugar, caramelizes at 160°C.

Science of Frying

Another common, albeit closely related, way to get your food deliciously brown and crispy is to deep-fry it. Now you might wonder how frying is different from sautéing, which is more or less what we have been talking about so far. But anyone who has had biryani will tell you that if not for the fried onions, it's not biryani. The difference is in the temperature of the oil. We sauté in the range of 110–150°C. Frying temperature starts at about 170°C. What happens when you drop a ball of thick urad dal batter into oil that has been heated to 170°C is described below.

The moisture on the outer layer of the batter will instantly evaporate and escape, dehydrating the surface to form a waterproof crust. This is what prevents subsequent dehydration of the insides of the vada, while still transferring heat to its innards. The outside then undergoes a rapid version of the Maillard reaction, browning quickly as the sugars and proteins in the dal combine to create a crisp, brown and delicious crust. As the insides heat up, the starches gelatinize, becoming soft and mushy. This is exactly how you get a crisp exterior and a soft, well-cooked interior when you deep-fry anything. This is why the vada batter must have as little water as possible. If it has too much water, the outside will take a long time to dehydrate, resulting in vadas that are overly dark brown with a thick crust.

An important thing to remember: You can't fry something that does not have both sugars and proteins. This is why you need a batter when frying something like chicken or pakoras. The batter is usually made using a starch, such as rice, wheat, corn or gram flour. You can also use breadcrumbs, which stick to whatever it is you are deep-frying, using eggs.

Of course, many things could go wrong. Let's take a puri. If the oil is not hot enough, the exterior will not dehydrate fast enough to become non-porous quickly. In that time, hot oil will seep into the puri and make it greasy and heavy. If the oil is too hot, the outer crust will brown too fast and the heat will not have enough time to cook the insides, giving you a raw doughy taste. Another thing to keep in mind is that hot oil oxidizes, and oxidized oil tastes nasty. This is why reusing frying oil is not recommended; however, if you filter for particles and prevent further oxidation by storing it in a dark place, you can use it again. One way in which you can reduce the amount of oxidation during frying is to use a narrower frying vessel, like the smallest kadai that fits your food. That will expose less of the hot oil to the air than a larger kadai. Also, fry in smaller batches. You can also add a pinch of baking soda to the batter of whatever it is you are frying. Alkaline conditions accelerate the Maillard reaction and will result in more even browning. Try it with fried chicken or vadas!

If you don't eat fried food right away, it will go limp. This happens because even after you take it out of the frying pan, it continues to lose moisture. If you don't provide an avenue for it to escape, it will settle on the surface of your fried chicken and make it soggy. The trick is to store freshly fried food at a temperature just short of the boiling point of water, so that no further cooking happens and it is easy for moisture to evaporate and keep things crisp. An oven at 93°C (200°F) will do the trick.

You can also take your deep-frying game to the next level by using something that tends to be taboo in the typical Indian kitchen—alcohol. But you'll have to wait till Chapter 5 for that.

4 Dropping Acid

When life gives you $C_6H_8O_7$, make $H_2O + C_{12}H_{22}O_{11} + C_6H_8O_7$

—Christopher Morley

Consider a cow. Remember Nandini? Let's picture her contentedly munching on one of the most versatile and successful life forms in the history of our planet—grass. After being artificially inseminated with prime semen from the Genghis Khan of bulls, she has just given birth to a healthy male calf and, thanks to the magic of genetic engineering, will produce close to 3000 litres of milk over the next one year. Her calf will require no more than 300 litres before being weaned, leaving the remaining 2700 litres to be consumed in your morning coffee, churned into butter, fermented into yoghurt, curdled into cheese and, in south India, poured over 100-foot-tall posters of film stars on the release of their next blockbuster.

But for now, let's consider yoghurt, or curd in the Indian context. In the West, curd refers to the solid part of milk after the liquid whey has been removed. To avoid confusion, we will use the term yoghurt (dahi in Hindi). To make yoghurt, you have to first heat milk to about 85°C.

This will kill all the bacteria in the milk, and there are a lot of them, and modify the proteins in it so that they don't separate into whey and crumbly curd when fermented. Think of this process as being akin to a homeowner evicting the nasty bachelors who are likely to make a mess out of his house and then leasing the house to a nice family of three that will keep it in in tip-top shape. The process kills the random assortment of natural bacteria and yeast that are likely to be chilling out in the milk and then introduce some of the bacteria that we know to be really nice guys. How do we know who the nice guys are? Just talk to your grandmother and she will likely tell you that she's known this bacterial family for generations. It is not uncommon for Indian homes to maintain some bacterial culture, which is done in a simple way. Take a tiny bit of yesterday's yoghurt and add it to boiled and cooled milk to make today's yoghurt. It all seems simple, but the real magic is in the chemistry of what happens when milk turns into yoghurt, and why yoghurt tastes the way it does—sour.

Introduction to Sourness

Among the five basic tastes (sweet, salty, bitter, savoury and sour), I think sourness is the most underrated. Even the word we use, 'sour', is not pleasant-sounding. It is, in fact, used to describe cats when they are annoyed, among other things. But if not for sourness, food would taste utterly bland and one-dimensional. An instinctively good cook understands that while salt and sweet enhance the flavours of your dish, sourness adds an entirely different dimension. That squeeze of lime into a simmering dal or salad does exactly this. Tamarind, for instance, is what makes a sambar a great sambar. Let me try and explain what I mean by adding a different dimension. If we talk in terms of music, salt would be like the volume knob. It makes everything louder and taste bigger. Try eating a pinch of cardamom powder. All you will taste is bitterness, while your nose will pick up on the volatile molecules in the spice. Now try eating it with a tiny pinch of salt. You will taste both the salt on your taste buds and smell the complex aromas of the spice. Salt amplifies other tastes. Sourness, however, is like the bass guitarist and mixing engineer combined. Take a boring singer and guitarist droning away, and then add a bass guitar to the mix. All of a sudden, it will feel like a full concert. The very addition of sourness makes the dish feel more complete than what it would have been otherwise. Sourness makes each ingredient stand out in its own place, which is what a mixing engineer does for music. Try drinking a strong cup of black coffee with a tiny squeeze of lemon. As odd as this might be, if you are someone who is sensitive to strongly bitter tastes, you will enjoy the coffee more (and trigger coffee Nazis, which is not a bad thing to do).

So, how do we perceive sourness? It's when an acid hits your taste buds. Well, not strong acids, since those will dissolve your tongue, but mild ones. Every sour ingredient is essentially an acid. In fact, the word acid literally comes from the Latin word *acidus*, which means 'sour'. To varying degrees of sourness, yoghurt, lemons, pineapples, tamarind, grapes and vinegar are culinary acids. Without them, food would taste one-dimensional.

Here's how sourness affects other flavours:

1. Sourness reduces bitterness: A squeeze of lemon into your spinach dishes will mute any residual bitterness in the cooked leaves. If you recall Chapter 1, leaves tend to turn bitter when cooked for too long. But be careful, acids can also decolourize green vegetables, so use them towards the end of the cooking process.

2. Sourness balances spice flavours by doing what an equalizer does in music. In dishes that have many strong flavours, adding

an acid will create space for every individual flavour and make them stand out.

3.   Sourness minimizes the perception of fattiness and makes food feel less heavy and rich. If you feel that you've used a bit too much oil in your dish, a good way to reduce the perception of greasiness is to add some acid (like lime juice) at the end.

4.   Sourness can balance overly sweet dishes. A good shrikhand is an example of this. If not for the sour tang of the yoghurt, it can be cloyingly sweet.

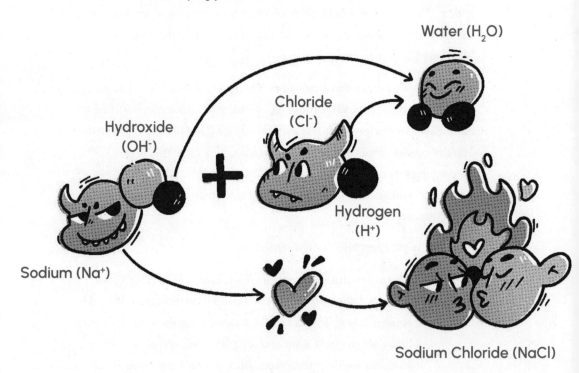

A base, sodium hydroxide, reacts with an acid, hydrochloric acid, to form sodium chloride (salt) and water. The positively charged sodium and the negatively charged chloride ions combine to form salt

As we have done in every chapter, it's time to zoom into the world of molecules and see what the deal with acids is. An acid is a substance that is capable of donating a proton to whoever wants it. A base is a substance capable of greedily grabbing and accepting protons from whoever is donating it. A simpler way to look at it is to consider the smallest and most abundant element in the universe: hydrogen. The hydrogen atom has one proton and one electron spinning around that proton. When this atom loses its electrons, as some atoms often tend to, it becomes positively charged. An acid is any substance that, when added to water, makes these hydrogen ions (single protons) available for any other molecule to use. When acids and bases react, they usually exchange protons in a sub-atomic drug deal and produce a salt and water.

I am, of course, oversimplifying, but for the purposes for understanding food chemistry all we need to know is that depending on the acid's sense of generosity, there are strong and weak acids and, correspondingly, strong and weak bases. The scale that measures this is called pH, where a value from 0–7 is acidic and 7–14 is basic. An exact 7 is, like the proverbial neutral centre of the taste universe, water.

The pH scale was invented by Søren Sørensen in 1909 while working for the Carlsberg brewery. The scale helped them produce consistent-tasting beer with a pH of around 4 (yes, beer is acidic).

What Is an Acid?

A pH value of 6 will be mildly acidic, while anything less than 2 will be strongly acidic. So, how does this translate into taste? The lower the pH, the sourer the taste. As one can see from the millions of enjoyable videos of babies being given a wedge of lemon to bite on, strong acids make our faces pucker. It is our brain asking us to be careful because not all sour things are good. Lime juice is fine, but an unripe fruit or spoilt milk can make you sick.

Remember my wife's hilarious episode, when she drank Coca-Cola to douse the effects of a really hot fish curry? Coca-Cola is so strongly acidic that its pH is lower than that of vinegar! If you are wondering how Coca-Cola then becomes palatable, given that trying to take a swig from a bottle of vinegar is not the most pleasant of experiences, remember our rules for sweetness from Chapter 2. Sweetness reduces the perception of sourness. Guess how much sugar you would need to make an acid as strong as Coca-Cola drinkable? A standard 330 g can of the drink needs a whopping 39 g of sugar, which is, in more visually impactful terms, nine full teaspoons.

Beyond adding sourness to dishes, acids play another crucial role in cooking. They denature proteins. Proteins are large, complex molecules that make up muscle fibres and connective tissues. When we cook meat, we use heat to denature proteins, causing them to become tougher. Another technique used with meat is marinating it in an acid. Any kebab is usually marinated in some acid, typically a mixture of yoghurt and lime juice. Now, the Internet (and many cookbooks) will tell you that acids make the meat more tender, but they don't! Acids break up protein structures and cause them to reassemble in bigger meshes, which actually makes the meat tougher. But in doing so, the acids allow the mesh to absorb other flavourful molecules in the marinade, like ginger, garlic and garam masala. By the way, all of this happens only on the surface of the meat. So, anyone telling you that the marinade penetrates deeper the longer you marinate is simply lying.

Marination attaches flavours to the outer layers of whatever it is that you are marinating. Another tip to marinating effectively: Beat the hell out of the meat with a tenderizing hammer, which will create more surface

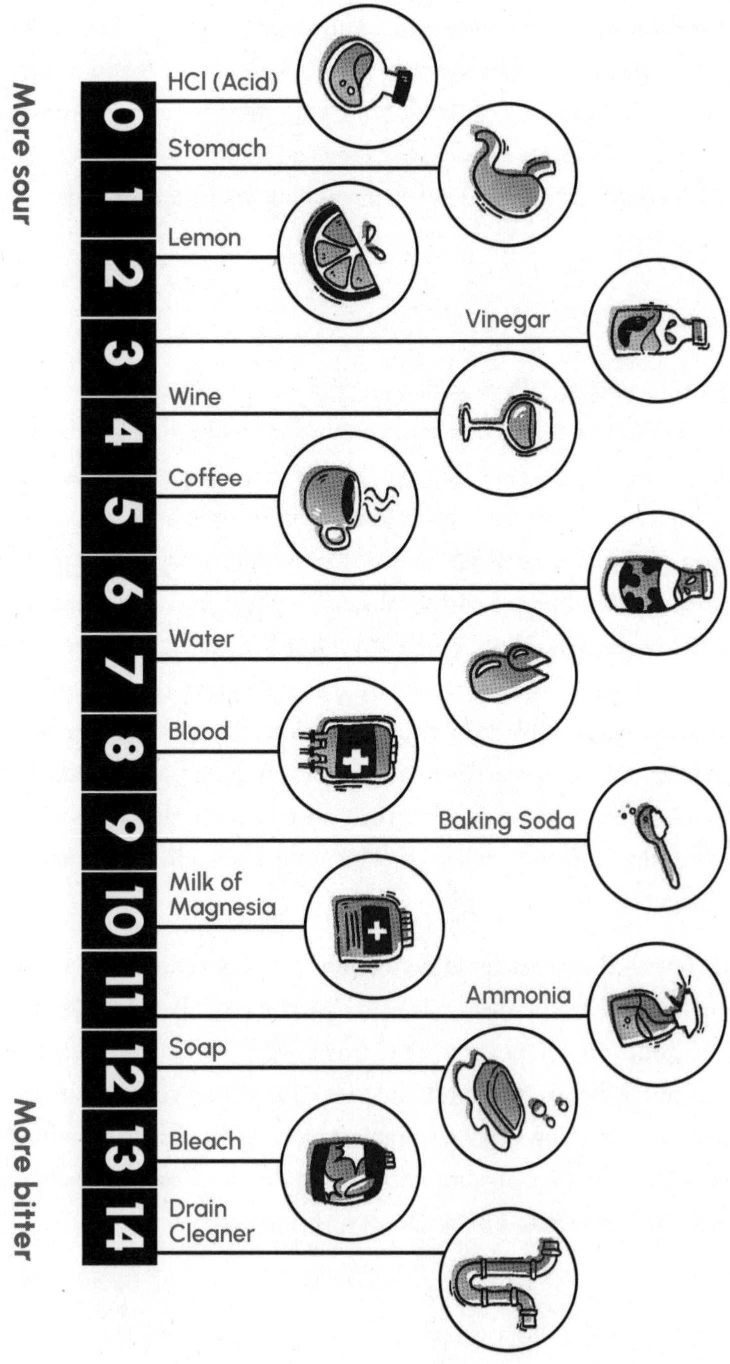
More sour
More bitter
0
1
2
3
4
5
6
7
8
9
10
11
12
13
14
HCl (Acid)
Stomach
Lemon
Vinegar
Wine
Coffee
Water
Blood
Baking Soda
Milk of Magnesia
Ammonia
Soap
Bleach
Drain Cleaner

area for the marinade to operate on. And use small pieces of meat. Avoid using over-acidic marinades, as they will cook the meat. If you want to take your kebabs to the next level, brine your meat for a few hours, and then marinate in spices and acid for about an hour. Twenty-four-hour marinades don't do much, other than cooking your meat in acid.

Science of Yoghurt

Yoghurt is a fantastically versatile, yet tricky, acid to use. The primary acid here is lactic acid, the same one produced by your muscles when they are tired. The advantage of using yoghurt is that you get both sourness and creaminess in one shot, unlike tamarind or lime juice. But if you overheat yoghurt, the delicate emulsion of milk fats, sugars and proteins will break up, leaving you with clumps of yoghurt that don't make for an appetizing dish. So, never overheat yoghurt. Always add it later in the cooking process. However, if you are making a gravy that is yoghurt-based, and need to cook it for a fair bit of time, the trick is to use some starch like corn flour, rice flour, wheat flour or gram flour (besan) and whisk it into the yoghurt to strengthen the emulsion, thus keeping it from breaking down when heated. This is the key to dishes like kadhi, morkozhambu and pulisheri.

Like it is with the batter for idlis and dosas, the convenience of picking up yoghurt at the supermarket is hard to argue with. But if you follow this method, you will never need to buy yoghurt again, as the one you make at home will be incredibly superior to anything you get at the store. Fermentation is a slow and deliberate process, and industrial production tends to use a ton of engineering tricks to speed it up. So, here's how you can make perfectly thick and delicious yoghurt at home.

1. Heat milk till the temperature is about 85–90°C. This is just short of boiling temperature. Do not let the milk boil. What this does is change the nature of the milk proteins, so that when they curdle, they set into large, solid curds and not separate into unappetizing crumbly bits. The higher the fat percentage in your milk, the thicker your yoghurt will be.

2. Switch off heat and bring the milk down to 45°C. This is the temperature at which the yoghurt bacteria are most productive. Try and use a vessel that retains heat better, like enamelled cast iron (pure cast iron might react with the acid in the yoghurt, so it's best avoided) or any thick steel vessel. This will ensure that the temperature does not drop too much during the fermentation process. Placing the vessel inside a switched-off oven also helps, as the temperature inside is not likely to swing a lot.

3. Once the milk is at 45°C, take a cup of that milk and add a teaspoon of yoghurt to it (the more yoghurt you add, the faster the fermentation. So, depending on the conditions in your home, you will need to experiment with the right amount), mix it in and then add this cup of milk back into the larger vessel. We do it this way to ensure that the yoghurt culture is more uniformly dispersed and fermentation rate is uniform.

4. Now place the lid and wait for four hours. As long as your vessel stays in the 40–45°C range, four hours is what it will take to set. At this point, taste your yoghurt (take a small pinch from the edge without disturbing the solid structure) and see if it feels right in terms of sourness and creaminess. If not, let it set for an hour more. The moment it tastes right, refrigerate it.

5. If you want to make hung yoghurt (like Greek yoghurt), just place your freshly made yoghurt on a muslin cloth and let the excess water drain out. You will then be left with a creamy, almost-spreadable, cheese-like yoghurt. Don't throw away the drained water. Add some asafoetida, salt and curry leaves to make a refreshing summer drink.

6. If you live in a place that is cold, you have to make some adjustments to this process. Because the temperature of the milk is likely to fall more rapidly than in summer, you need to add more yoghurt culture (double the usual amount) to give it a better head start. You can then place the vessel inside your microwave with the light on and then wait for anywhere from 8–16 hours. It will eventually ferment, so don't give up too soon. An easy technological solution to this problem is to buy an electronic pressure cooker (like the Insant Pot). Most of them have a 'yoghurt' setting that keeps your milk vessel precisely at 45°C for the entire duration and will yield you perfectly creamy and thick yoghurt in under 4 hours every single time. Chapter 6 discusses electronic pressure cookers in detail.

Science of Tamarind

The word tamarind comes from *tamar-e-Hind* (dates from India), despite the fact that it originated in Africa, because that's simply where the Arabs assumed it came from. This fantastically sour pulp is an absolute staple in almost every kitchen in India. It lasts for ever and also survives long cook times (unlike lime juice). The primary acid at work here is tartaric acid. There are many variants with different flavour profiles ranging from mild to strong sourness. The typical way to use it is by soaking it in warm water for 10–15 minutes and squeezing out as much of the juice as you can before discarding the fibrous bits.

There is also kokum and Malabar tamarind, but these aren't related to the tamarind botanically and have a milder flavour profile. Also, these are not as fibrous and do not require soaking and filtering. You can just drop them straight into dishes to add some sourness. They tend to be preferred in more subtle, seafood-based gravies, where the overwhelming sour punch of tamarind is not preferred.

A doubt that a lot of new cooks tend to have is this rather common, yet highly dubious, instruction in most recipes on the Internet: Cook the tamarind juice till its raw smell goes away. What does 'raw' mean here? A more sensible way to ensure that the tamarind does not overwhelm your dish is to keep tasting it till it has the level of sourness that you think is acceptable. In general, tamarind water-based gravies will need 6–8 minutes of medium heat to bring the sourness down to an acceptable level. Start from there and then decide if you need less or more cook time. Contemporary chefs who work with food scientists have determined that the pH of a good, balanced dish tends to be in the range of 4.3 to 4.9. So, while I am not recommending that you invest in a pH meter to test your sambar, this is a reaffirmation of the basic idea that a perceptible level of sourness is critical for most dishes. As we learnt before, a pH value of less than 7 is acidic.

Tamarind's high levels of acidity also make it a useful way to clean metals prone to oxidization (silver, copper and copper alloys such as brass). If you are rich enough to own copper pots, using tamarind and salt (as an abrasive agent) to keep them shining is not a bad idea.

Science of Mango

Mangoes, whose sweeter and riper varieties tend to hog all the limelight, play a big role as a souring agent in Indian cuisine. When they are in season, raw mangoes are used to add a fruity, sour flavour to a wide range of dishes like avial and fish curry in the south and dal in the west. When out of season, sun-dried, unripe mangoes are powdered into amchoor, which serves as a souring agent for a lot of dishes, particularly in the north. It has the advantage of being able to add sourness without adding moisture, which is why it is the best choice when it comes to snacks and dry dishes. It can also be used a tenderizer when marinating meat. You can achieve a more layered sourness in your meat by using a combination of yoghurt, lime juice and amchoor, instead of just using one acid.

Chaat is a great example of building an intense disco-party-in-the-mouth flavour by layering multiple acids. The first layer of acid is the chopped tomatoes, which are mildly acidic while adding umami (more on this in Chapter 5). Then we have the tamarind chutney, which provides the bulk of the base sourness. On top of this, we have the green chutney, which has lime juice, providing the top layer of fresh-tasting acidity. Finally, the chaat-wallah also sprinkles some amchoor at the end to provide a bridge between the fresh, citrusy zing of the lime juice and the strong base flavour of tamarind. Chaat is a great lesson for home cooks in how to make your food interesting. If a recipe calls for amchoor alone, try using half the quantity mentioned and squeeze lime at the end to bring about a more layered and richer flavour profile.

It's important to understand the flavour profiles of different acids, as we did with spices. Tamarind has a meaty, savoury base flavour that works well when cooked, sort of like a drummer and bass guitarist put together. Amchoor has a fruity, bright, sour flavour that works best when not cooked too much, like a lead guitarist. Citrus juices, which we shall discuss now, are like the lead vocalists who are usually in your face.

Science of Citrus Juices

Sometimes when your dish is missing sourness, and you realize it only at the end, adding tamarind or amchoor may not work. Tamarind is too overpowering and requires cooking, and last-minute amchoor works for chaat but not for a more delicately flavoured dal. This is because, in addition to the sourness, amchoor will lend the flavour of mango to your dish. While this might not sound like a bad idea, it may not be something you want either. In such situations, the best way to add sourness, without adding additional flavours, is to opt for citrus juices. There is a reason for this. When we add tamarind, we add the entire fruit to the dish, which gives us all of its complex flavours. Likewise, amchoor is just dehydrated mango that has all the flavour molecules of the fruit, in addition to the sourness. When we use lime juice, which is mostly citric acid ($C_6H_8O_7$), we avoid the pulp and peel of the fruit. And while it is the peel that has all the essential oils of citrus, which give it its unique flavour, we don't want them because they will overwhelm our dish. Not unless we are making pickle or a lime-based dessert. However, when we squeeze lime, a tiny portion of the essential oils does get into the dish, but it is just enough to provide a hint of citrus without overpowering the dish. By the way, the seeds are really bitter, so avoid them at all costs. The white part under the skin doesn't taste great either and is best avoided.

A key thing to remember about citrus is to never add it early in the cooking process. Cooked citrus juices develop strange, nasty, bitter flavours. This is why they are added right at the end.

If you want to make things interesting, try different kinds of citrus juices—orange, sweet lime, citron, etc. And if you want to take things to the next level, squeeze limes and let them sit at room temperature for a few hours. The terpenes in it (Chapter 2) will get oxidized, which will then develop a well-rounded taste. So, if you are making nimbu paani, lemon rice or jal jeera, squeeze the limes well ahead of time. If you are using it just as an acid, don't bother. Just squeeze it straight into the dish. Also, don't try this trick with oranges—they will taste off if exposed to air for long.

Science of Vinegar

Vinegar, from the French word for 'sour wine', comes from the oxidation (which, in simple terms, is oxygen being greedy as hell for other atoms' electrons) of wine, aided by a bacteria named Acetobacter aceti, which despite sounding like the opening lines of a verse from the Upanishads, has the ability to convert ethanol to acetic acid in the presence of oxygen. The simplest vinegar is distilled white vinegar, which is pure acetic acid diluted with water. It's pretty cheap and one of the few chemicals (baking soda being the other one) that does a stellar job both in the cooking and cleaning department. It is a fantastic culinary acid that can be used to pickle vegetables, marinate meat and, in combination with fat, dress salads.

Vinegars involve a two-step process, of which the first is fermenting something sweet into alcohol. This could be wine made from grapes, or any other alcoholic beverage made from a fruit. Once you have that, introduce our Upanishadic bacteria, Acetobacter aceti, into the mix and let it turn the alcohol into acetic acid. Red wine vinegar will, in addition to the sourness from the acetic acid, have the flavour molecules of red wine, just like apple cider vinegar will have molecules of fermented apple juice. Fruit-based vinegars have a more interesting taste than plain white distilled vinegar. Unfortunately, vinegars are under-utilized in Indian cuisine, except in Goan cuisine. This could be because of the historical love-hate relationship the subcontinent has had with alcohol in general, and you can't make vinegar without making alcohol first.

A great way to experiment with vinegars is to replace tamarind or lime juice with it. Use about half the amount of vinegar as you would of lime juice. Red wine vinegar works well in meat-based dishes, where just a touch of sourness is required, while apple cider vinegar works really well in chutneys (instead of lime juice). You can also pickle (peeled) boiled eggs in a solution of salt, sugar and apple cider vinegar. You may also add some green chillies into vinegar and let it steep for a few weeks before filtering out the chillies. This will give you a vinegar that will add both heat and sourness to your dishes. You can even steep roasted whole spices in vinegar for some garam masala vinegar.

Science of Tomatoes

Tomatoes, as acids go, are tremendously temperamental. Depending on the season, they are likely to be sweet and fruity, or tasteless, or even mildly tart. In fact, every trip to the grocer is likely to yield tomatoes presenting a wide range of sourness. Green, unripe tomatoes are more predictably tart, but they are not always available.

I was in a dilemma over whether or not to include this juicy member of the nightshade family in the chapter on acids, or to give this original native of Peru a chapter of its own. Like onions, tomatoes are central to most regional cuisines in India, with a large number of gravies built on a foundation of onions and tomatoes. In fact, for centuries after the Europeans discovered it in the Americas, tomatoes were considered to be poisonous. This was because they used dining plates made of pewter, which has a high amount of lead. If you recall Chapter 1, the reason we don't use cast iron vessels when cooking tomatoes (or any other acidic ingredient) is that the acid will leach the metal. This is exactly what was happening there. The tomatoes would leach lead from the pewter plates, slowly killing several aristocrats off from lead poisoning.

On that macabre note, let us get down to the science of the tomato. Most of the flavour is in the gooey pulp and the seeds, so please throw away any recipe that asks you to discard them. The pulp is particularly rich in glutamates (more on this in Chapter 5), which is what adds the umami flavour to any dish featuring tomatoes. The flavour of tomatoes improves with cooking and concentrates when you sun-dry or dehydrate them. When recipes call for tomato puree, noobs add tomato puree, experts add tomato paste and legends add tomato ketchup. Tomato paste is highly concentrated and provides a more consistent and rich sourness than the general flavour lottery that comes with fresh tomatoes. But tomato ketchup, which also has onion powder, garlic powder and vinegar, is the secret weapon of the expert cook. When a recipe calls for tomatoes, add the fresh ones, and then drop in a sachet of the tomato ketchup that you should be saving up from all your home deliveries. Ketchup will improve any red-coloured gravy. For special occasions, especially when making gravies where tomato is the star, say a paneer makhani, you can apply the flavour-layering principles outlined so far. Use fresh tomatoes, tomato paste (or ketchup) and dehydrated tomato powder to get the most intense, layered flavour.

The flavour of tomatoes is improved significantly by concentration (removal of water) and sustained low heat over long periods of time. The longer you cook tomatoes, the better they taste, which is why some pasta sauces are cooked for an entire day in Italy. Given that Indian cooking adds a ton of other flavouring ingredients, we won't necessarily notice the difference. But when you make something like a makhani gravy in bulk, try and cook it low and slow to see the difference.

When using tomatoes in salads, salt them ahead of time. Not only will the salt pull out some of the moisture, resulting in a more concentrated flavour, a salted tomato will make you eat more salads because it is absolutely delicious. For some reason, most restaurants in India serve unsalted tomatoes in every single salad, which is a pity.

Other Culinary Acids

Here is a list of a few more culinary acids in the Indian kitchen:

1. Dried pomegranate seeds (anardana) is another way to add an intense, complex and fruity sourness to dishes. It is used in certain Punjabi dishes, such as chana masala, and in chutneys.

2. CO_2 in water (essentially soda, but not to be confused with baking soda) is also acidic, which is why food consumed with soda tastes better than plain water. Acids activate salivation.

3. Alcoholic drinks are also acidic, which is why cooking with wine or beer is common in the West. But given the general tendency to keep alcohol at an arm's length in Indian homes, this technique is not very common in the country. If you have some old wine lying around, try this when making gravies: After you sauté the spices in oil, cook onions, ginger and garlic, add a splash of the wine into the pan. This will deglaze all the lovely tasting brown bits (Maillard reaction, remember), extract more flavour from the spices and, while at it, add sourness.

At this point, you might wonder why we tend to prefer acids over bases in our food. If you recall, a bunch of expert chefs in the West had got food scientists to measure the pH of their best dishes, all of which were in the range of 4.3 to 4.9, which is moderately acidic. There is an evolutionary explanation that begins with our good old photosynthesizing magicians, plants. Plants literally turn thin air (CO_2) and sunlight into the mass that constitutes it, and that is a lot of hard work. Over billions of years, another family of annoying living things called animals came along and, instead of tapping into the sun's energy directly, started munching on plants as a shortcut. This gave them a massive advantage—the ability to have a high rate of metabolism—because you can get a ton of energy by eating a potato in a few minutes, while the plant took several months to make it in the first place. This advantage allowed animals to ultimately move around.

But the plants didn't suffer quietly while this marauding horde of mobile critters literally stole the fruits of their labour. They developed defence mechanisms to prevent such uncontrolled consumption. One of the rather effective things they did was to invent a family of molecules called alkaloids, which tend to be poisonous for animals. That is not to say that the animals sat around munching on poison and dropping dead in large numbers; they figured out ways of detecting poisonous plants before ingesting them, which helped us hone our perception of bitter tastes. Since alkaloids tend to be bitter, our tongues evolved a mechanism to detect these (at the back of the mouth). This mechanism causes a nasty aftertaste to linger in our mouths, causing us to spit out what we are eating, potentially saving us a gruesome death by poison.

All things alkaline, which by the way are not related to alkaloids, tend to taste bitter, in that they activate the same receptors. So, over millions of years of evolution, our digestive systems have developed a bias towards acidic foods. It's important to note that while most poisons tend to be bitter, not all bitter things are poisonous. And alkaloids aren't just dangerous poisons. Human ingenuity has turned them into life-saving drugs poisonous enough to kill the parasites that cause malaria and tuberculosis, without affecting the human. Without alkaloids, we would not have a pharmaceutical industry.

But back to our acids for now. If you want to make things interesting in your kitchen, try out different kinds of acids. Take recipes that use tamarind and replace it with vinegar instead. Always remember, it is best not to add citrus when cooking, but you can add vinegar. Or you could try raw mango, or kokum. In dishes that call for a squeeze of lime, try squeezing orange or pineapple juice. Interestingly, pineapple juice makes for a fantastic marinade because the bromelain molecule is particularly good at tenderizing meat, while the sugars enhance the flavour of whatever it is that you are marinating, in addition to encouraging more browning reactions. Green apples also make for an interesting way to add sourness.

Alternatively, if you just want a sour taste, without the sweetness or bitterness of the other components of these fruit juices, you can go straight to the source and directly add powdered tartaric acid (from grapes), malic acid (from apples) or citric acid (from citrus fruits).

Another interesting, albeit rarely used, acid is tea. As we saw in the opening chapter, dropping a teabag into the pressure cooker when cooking chickpeas is a fantastic way to neutralize any unused baking soda, which is basic and adds a bitter aftertaste. The added bonus is that the tea will impart a lovely dark brown colour to the chana. Remember Chapter 2, where we discussed how visual appeal plays a role in flavour perception? Foods that are a darker brown in colour tend to indicate more delicious tastes because of the Maillard reaction. This is why darker-looking chana, despite not actually undergoing Maillard browning, appears to be more flavourful than it actually is.

The Acid Cheat Sheet

Here's a simple guide for you:

1. Acids add brightness to food, while balancing other flavours, particularly spices.

2. Acids cause us to salivate. The processed food industry takes advantage of this weakness of ours to make what is nutritionally terrible food taste delicious. And since adding acids balances the saltiness and sweetness, it allows them to cram more salt and sugar than is good for us into the snacks. This is not to deny that they taste delicious as a result.

3. Acids tenderize meat, in that they break down the protein structures that help flavour molecules attach themselves to the surface. However, strong-enough acids toughen meat, so use them in moderation.

4. Acids also cause plant cell walls to toughen up by bonding with the pectin, which is why cooking lentils with acids (such as tomatoes or tamarind) takes longer.

5. A good way to use acids is to layer them, as you might do with spices (recall Chapter 2). In dishes like fish curry, sambar or kadhi, the acid is the anchor, while in dishes like dal, acids are the accent on top. You can layer acids by using different ones at various stages of the cooking process. Chaat, as described earlier, is a fantastic exemplar of acid layering—tamarind and tomatoes act as the base, while amchoor and lime juice, and occasionally pomegranates, are the accents. And remember what heat (capsaicin) does to flavour perception. Chaat, being hot, also results in an endorphin rush that makes the overall eating experience more pleasurable.

6. Coca-Cola is more acidic than vinegar. As much as nine teaspoons of sugar in one can of the beverage is what makes it palatable. If you add so much sugar to vinegar, it will taste pretty decent too.

7. Liquid acids preserve anything stored in them because bacteria don't like living in acidic conditions. Interestingly, some microorganisms use this as a trick to keep competing critters out. Yeast and yoghurt bacteria produce lactic acid, which makes the fermented product too acidic for other fungi and bacteria. This is why fermentation works. If it was a free-for-all for every microbe out there, it would be called spoiling and not fermentation.

You can take a good dish and add an acid to make it an amazing dish.

5 Umami, Soda, Rum

*I enjoy cooking with wine. Sometimes
I even put it in the food I'm cooking.*

—*Julia Child*

The first half of this book focused on things most cooks know intuitively, such as how heat works, how pressure-cooking works, how flavours are extracted from spices, the Maillard reaction and the role of acids. The focus was on explaining the science behind why something you do in the kitchen works. If you understand how what you do in the kitchen produces a certain outcome with a specific ingredient or technique, it empowers you to apply that knowledge more widely across other dishes

The second half of this book will shift the focus to things that are more commonly misunderstood and tend to be victims of pseudoscience.

In 1908, Kikunae Ikeda, a chemist and professor at the Tokyo Imperial University, was trying to figure out why the broth his wife used to make soup tasted so much better than the others. After presumably ruling out the bias arising out of heartfelt love for her, he figured that it was the seaweed (kombu) that made all the difference. A broth made with seaweed had a meaty, savoury and lingering taste that coated the tongue, while a broth made without it was distinctly underwhelming in comparison. A chemical analysis of the seaweed revealed that the magic molecules were salts of glutamic acid, an amino acid that is one of the building blocks for proteins.

Subsequently, a student of his discovered two more molecules that had this coat-the-tongue-and-play-lingering-crescendo-notes property. These were guanosine monophosphate (GMP) and inosine monophosphate (IMP). Like it is with other tastes, our tongues have receptors into which one part of these molecules fits snugly and activates them. It turns out that combining glutamates with IMP or GMP has a greater-than-sum-of-parts effect. Essentially, adding both will result in an intensity of flavour that is a lot more than what you might expect by adding them individually. This is something cooks have known intuitively around the world. The Japanese combine kombu (that has glutamates) with bonito flakes (that are made from tuna and have IMP) to make dashi stock, which forms the base for a lot of Japanese dishes. The Italians combine parmesan cheese and tomatoes, which are also quite rich in glutamates. The Japanese, in particular, have a cuisine centred around umami, which amplifies all other flavours, whereas Indian and Middle-Eastern cuisines are centred around fats and oils, which primarily transport flavour molecules. This is also why Japanese cooking is rather minimalistic. With umami as the base, you can make dishes using very few ingredients and yet get tremendous depth of flavour. Umami makes other flavours come a long way.

This brings us to a question: Why is it that Indian cuisine seems to not pay too much attention to umami? You might be tempted to think that this is because a lot of people in the subcontinent wrongly believe that the sodium salt of glutamic acid is poisonous and can cause brain damage in children. This is not true. If you examine the past, rural communities in India have used umami-heavy ingredients and pairings for a long time. Most fermented food, particularly which involves grains and fish, is extremely rich in umami. The standard pairing of onions and mutton, in the form of bhuna gosht, is yet another classic glutamate-inosinate pairing. Fisherfolk in India use a lot of anchovies. To take a slight detour into social awareness, the ubiquity of anchovies among coastal communities has to do with the fact that they cannot afford the richer, fattier, larger fishes, the ones that are sold to middle-class folks. The joke is on us here because it's the tinier fish that pack the most amount of umami. The cuisine of the North-East is also rather umami-heavy, thanks to the use of meat stocks and cabbage, which is also relatively high in glutamates.

Monosodium glutamate (MSG), is umami in its purest form. There is no chemical difference between the glutamates in cabbage, mushrooms and tomato, and the half teaspoon of Ajinomoto that you sprinkle on fried rice. If you think MSG is bad for you, you might want to let every nursing mother know. After all, breast milk is exceedingly high in glutamates (at least ten times the amount of glutamates than regular cow milk). And if even that doesn't convince you, there is approximately 2 kg of glutamates in the proteins that make up our body. A teaspoon more of that stuff will not kill you. That said, a really tiny percentage of people are allergic to MSG and must avoid it. But this is worse than the gluten-free fad. The number of people who have coeliac disease, which causes an allergy to gluten (meaning they cannot eat wheat) is far more than those who are actually allergic to MSG. Yet, the number of people who avoid MSG is a lot more than those who avoid gluten. So, it is safe to say that MSG is likely safe when consumed in moderation, a truism that is more or less true for all kinds of food.

The Fifth Taste

It's now time to dive deep and understand how umami works at the molecular level. Our tongues have T1R1/T1R3 receptors that are activated in the presence of glutamates. If the food also has IMP, or GMP, then the glutamates get even better at keeping these receptors on. It's interesting to note that both IMP and GMP are made of the same material that makes up DNA and RNA, which means they are present in all living cells. What this taste feels like is an intense, savoury, lingering feeling that amplifies all other flavours. Since we know that glutamates work even better in the presence of IMP and GMP, always pair IMP- and GMP-heavy ingredients, such as meat and mushrooms, with glutamate-rich ingredients like tomatoes or onions for an even more amaklamatic umami-bomb flavour.

There is still a little bit of mystique surrounding umami. A lot of research on how umami works and why we have a taste dedicated to it is still underway, and we are learning new things all the time. For instance, we now know that in addition to having a separate taste receptor for umami, it looks like glutamates (and its compatriot nucleotides IMP and GMP) may be causing a phenomenon where other taste receptors, particularly salt and sweet, stay activated for longer periods of time than they would have in the absence of glutamates. This explains the lingering aspect of umami, where it causes other strong tastes to last longer on the tongue. Since umami amplifies saltiness and sweetness, a good way to reduce your salt and sugar intake is to use MSG, which, in addition to adding umami, is only about one-third as salty as common salt. If you remember Chapter 2, saltiness is typically the tongue's ability to detect sodium in food.

The reasons why umami is classified as a separate taste are still open for discussion. One theory being explored is that it evolved as a mechanism for the tongue to detect protein-rich foods that are healthy. Meat, compared to plants that either want you to eat them (fruits) or want

you to not eat them (spices and vegetables), tends to be low on strong flavours. So, umami might have evolved as a way to detect and develop a taste for proteins. The fact that we can detect glutamates even at lower levels of concentration (up to six times lower than salt) lends some weight to this explanation.

The Magic of Baking Soda

Picture the Grand Canyon. A gaping 1.6 km-deep hole carved out by a meandering river that had little else to do over millions of years. Now imagine a situation where there is a cheap, simple household substance, but where the gap between its magical abilities in the kitchen and the Indian public's perception of it is as large as, to quote John Oliver, 'That part of the state of Arizona where there is distinctly less of Arizona.'

This substance is sodium bicarbonate ($NaHCO_3$), also known as baking soda or cooking soda. It's also one of the main ingredients in fruit salt, also known as Eno, which is a mix of sodium bicarbonate, sodium carbonate and citric acid. It is commonly used as an antacid. There are many people in India who will refuse to use baking soda, but opt for fruit salt when making cakes because baking soda has the worst PR agency in the world and needs to hire me right away. This chapter is my pitch to baking soda. Please hire me and I will fly down in a helicopter to the bottom of the Grand Canyon, throw down a rope and elevate you 1600 m, to the top of the canyon where you belong.

First, let me get the most common misconception out of the way. Baking soda in small quantities is not bad for you. In large quantities, everything is bad for you. Thanks and regards. Now, let's move on.

Let's begin with what it can do to pectin. Baking soda is the guy operating the wrecking ball on pectin. Adding a pinch of baking soda when cooking legumes like chana, rajma and black urad dal will reduce the cook time and fuel consumption by about 40 per cent. Adding a teabag to the cooker will ensure that any unused baking soda is neutralized. Because baking soda is mildly basic, it has a soapy and bitter taste, so the idea is to add just enough quantity for it to do its job but not linger around unutilized.

This trick is useful in several other situations too. For instance, you can put this to use when you want to make the most amaklamatically crispy potatoes in India, which is not an easy thing. Potatoes in India tend to be low-starch, high-moisture varieties, so they don't get crisp easily. The outer layer burns well before enough moisture has escaped to make it crisp. Enter our hero on a helicopter from the floor of the Grand Canyon, armed with a Gatling gun and sporting Ray-Ban shades. A pinch of baking soda in the water you boil peeled potatoes in will break down the pectin, resulting in rough, jagged surfaces with significantly more surface area for crisping. Next, sauté these potatoes to get the most amazingly crunchy texture and golden brown colour. You can also use a pinch of baking soda while blanching green vegetables. This will keep the vegetables green, as the baking soda will prevent the breakdown of chlorophyll, which gives the vegetables their characteristic colour. Don't cook them for too long though because the baking soda's assault on pectin can turn your vegetables to mush.

Baking soda can also tenderize tough cuts of meat. If you recall Chapter 1, a common misconception is that using acids in a marinade helps make the meat tender. They do not. On the contrary, acids make meat tougher. Bases, on the other hand, can make it tender. If you add a pinch of baking soda to tough cuts of meat, like beef or mutton, and let it sit for 5 minutes, it will make the meat tender. But don't add too much or you will be left with a nasty aftertaste.

We aren't done yet. Baking soda has the ability to accelerate the Maillard reaction, the one that turns ordinary food deliciously brown. Anytime you want more browning, sodium bicarbonate is your friend. A pinch added to vada or dosa batter will produce restaurant-grade dosas and vadas (now you know what they are doing). A pinch added when sautéing chicken will give you the right amount of browning before the chicken dries out.

Have you noticed how we have been talking only about the non-baking uses of something named 'baking' soda. Yes, that's how much of a renaissance person this molecule is. If you bring an acid to the party, baking soda will be happy to produce carbon dioxide to help bake cakes and bread, and leaven idli, appam and dosa batters in case you are doubtful about the quality of the job done by the bacteria. You can even make a fluffier omelette by adding a pinch of this to the eggs before cooking them.

Once you are done making the crispiest of potatoes, pressure-cooking the softest chana and baking a loaf that rises like the LIC building in Chennai, you can use our hero to clean your kitchen. It's an effective abrasive, and combined with vinegar, one of the best jugaads for cleaning. Baking soda also absorbs musty odours, so a cup of it in the fridge will significantly keep strong smells from settling into other food items.

The magic of Baking Soda

Helps in removal of strong odours

Breaks the cell walls of pulses and legumes like Dal and Rajma, helping them cook faster

Used for relief from acid reflux

Helps caramelize onions faster

When added to ba[ter?] it helps with even browning when fry[ing]

Can be used in combination with vinegar as a cleaning agent for pots and pans

Can be used to bake cakes or bread when combined with an acid like yoghurt

$NaHCO_3$

The Magic of Alcohol

The subcontinent has had a love-hate relationship with alcohol. A sizeable percentage of the population does not touch it in any form as a result of religious strictures, while a portion of the remaining number regularly stands in lines at state-run shops that usually monopolize the sale of spirits. The sizeable tax revenues stemming from them make up most of individual states' budgets thanks to a fundamentally flawed federal system that centralizes all other tax revenues while rewarding poor administrative behaviour. Some states, however, have prohibition, which is when the black market steps in and does a stellar job of catering to the tipplers' needs. And then there are states that have strong voting blocks of women who will oppose any relaxation in the availability of alcohol. The complexity of this part of the world, which stems from structural poverty and the entrenchment of a caste system, where the British and upper-caste communities worked together in the past to snub rich, local brewing traditions that made kallu, feni and a thousand other fermented spirits, is not a topic for this book. Alcohol is politics in this part of the world, and there are no easy answers. As an urban, well-to-do hipster, my whining that I don't get craft Belgian beer in the state-run Tamil Nadu State Marketing Corporation (TASMAC) store near my house flies in the face of a complex large-scale issue involving alcoholism and its concomitant evil, domestic violence. That said, armchair-uncleji theories about how tropical parts of the world do not have brewing traditions because of their climate are patently silly. When you put human beings, carbohydrates and some microbes together, alcoholic drinks will emerge. This brings us to a useful segue in our chapter. When you ferment things, ethanol is almost always produced. There is no escaping it.

If you make yoghurt at home, it will have a tiny amount of ethanol that is produced by wild yeast in the environment. Industrially produced yoghurt is fermented in sterile conditions to ensure that only specific strains of bacteria, such as streptococcus thermophilus and lactobacillus lactis, feast on the milk sugars and do not produce alcohol. This is also why homemade yoghurt is almost always richer in flavour, because a diversity of microbes results in more complex flavours. And some alcohol.

If you bake bread, there is literally no escaping alcohol production. Yeast produces alcohol when it eats up sugars. In fact, one element of the sweet smell of baking bread is the ethanol evaporating. The alcohol molecule has a slight resemblance to the sugar molecule, which is why, in low concentrations, it tastes sweet. Breads, such as naan, made after a long fermentation process with yeast, can sometimes contain about 2 per cent residual alcohol even after baking. Leave aside bread-bakers and naan-makers, the long-standing tradition of eating rice fermented overnight involves consuming a finished product that can be pretty alcoholic. Now you know why people eat fermented rice with raw onions soaked in yoghurt in south India. The onions overpower the smell of alcohol.

If you, like most frugal Indians, have no problem eating slightly overripe fruit instead of throwing it away, please remember that the process of ripening, beyond the point of the perfectly ripe fruit, is largely the process of fermentation by bacteria and yeast. A very ripe banana can have up to 0.5 per cent alcohol by volume. Another alcohol-containing ingredient in the kitchen is soy sauce, which too is a product of fermentation.

If you are wondering where I'm steering this conversation, here's the thing. There are three positions adopted towards alcohol in this part of the world. The first one is a religious proscription that bans the use of alcohol in any form. The second one is to do with a tiny segment of the urban population that makes al dente pasta with San Marzano tomatoes and white wine from the Lombardy region. For now, we shall ignore these two categories of people and focus on the third.

If you are someone who answers yes to any one of these questions:

1. I generally do not drink alcohol, but it's not a religious thing, just a personal choice.

2. I occasionally drink beer or wine, but nothing else.

3. I do not have alcohol intolerance (where your body cannot break down alcohol).

I want you to consider using alcohol in moderate amounts while cooking. You don't have to drink it, and rest assured, your finished product will not be any more alcoholic than the bread you bake or the yoghurt you ferment.

Let's briefly go back to our chapter on spices and flavour. If you remember, flavour is a multisensory experience that involves the taste buds, nose, ears, eyes and mouthfeel. Among these, aroma contributes to about 80 per cent of how we perceive flavour. While our tongues can detect five kinds of tastes, our noses can detect thousands of aromas. In fact, we can detect some aromas even if the concentration is of a few molecules in a trillion! Sulphur-based aroma molecules, like the smell of cooked fish, tend to be detectable at concentration levels approaching one molecule in a quadrillion!

This is a crucial thing to remember when making good food—we can only taste things that are water-soluble, but we can smell way more volatile aroma molecules thanks to the olfactory receptors in our noses. Most spices and strongly aromatic ingredients have volatile flavour molecules that are not water-soluble, so we can't actually taste them. Remember, you smell cardamom, you don't taste it. What you taste when you bite into cardamom is its woody mouthfeel and bitter taste. This is why fats are absolutely crucial to cooking, because most flavour molecules are fat-soluble, not water-soluble. This means that when you cook spices in hot oil, it extracts all of these flavour molecules and dissolves them into the oil, thus preventing them from being lost to the air. When you eat this food, the enzymes in your saliva start breaking down the fats, which results in those dissolved aroma molecules escaping into your mouth. As they enter the short, shared highway that transports both food and air, the act of breathing out elevates the aroma molecules, which are basically gases, and makes them hit the olfactory receptors. That is when you truly experience the complex taste of the thousands of aroma molecules from the saffron in your biryani.

Now that I have your attention with the words 'saffron' and 'biryani' (no political angle here, I assure you), let me tell you that after fats, the silver medal winner in the 100 m flavour extraction race is alcohol. A tiny amount of alcohol used while cooking will almost always result in a stronger flavour. The alcohol will help transport more aroma molecules to your nose and make your dish pop.

Here is how I use alcohol when making Indian dishes:

1. A splash of vodka, brandy or rum when cooking onions, ginger, garlic, tomatoes and spice powders has two benefits: extraction of more flavour from the spices and the alcohol's ability to release all those sticky bits from the bottom of the pan, which have a ton of flavour thanks to the Maillard reaction.

2. A splash of wine added at the end of a dish, along with finishing spices, will amplify the effect of those spices when you eat.

3. Keep in mind that while a small amount of alcohol can amplify flavour, a large amount will actually prevent the release of flavour molecules by holding on to them like family heirlooms. This is incidentally one of the reasons why bar snacks tend to be overpoweringly spicy in India. When had

with a large whisky, that mirchi bajji could well be made using bhut jolokia chillies and you won't notice.

4. The amount of alcohol in beer is not strong enough to make a difference, so at the very least, use wine. The cheapest one will do because once heated, all the evocative notes of strawberries and smoked salmon in your fine Chardonnay will largely be destroyed. You can, however, use beer as an acid (see Chapter 4).

5. If you are frying fish (or vegetables) using a batter, try a batter made of maida, salt and vodka (which is just plain ethanol diluted in water). What the alcohol does is reduce gluten development, which we do not want in a fried product, as it will cause chewiness. It also prevents surface starches from absorbing too much water to gelatinize, which will result in a drier and crispier crust. This technique was pioneered by Heston Blumenthal, who went one step further and carbonated his batter before use. The aerated batter makes the crust airy, in addition to being crisp. Trust me, use this technique and you will have some game-changing pakoras to enjoy.

6 Taking It to the Next Level

In the strictest scientific sense, we all feed on death, even vegetarians.

—*Spock*

Science of Microwaves

Most urban, middle-class households have a microwave oven, but a majority of them use it for little more than reheating food. This is a pity because a microwave oven is a versatile device that can, if not necessarily transform your cooking, significantly save you time and mess in the kitchen. As always, we shall first focus on the physics of it.

Microwaves are electromagnetic waves, no different from visible light or radio waves. Did you notice how I framed that sentence slightly differently from how you are probably used to hearing it? Had the sentence said 'microwaves are electromagnetic waves similar to cancer-causing X-rays, gamma rays and UV rays', your perception would have been different. Well, not surprisingly, microwaves are closer to visible light and radio waves than they are to UV rays, X-rays or gamma rays, all of which have more energy than visible light.

169

How a microwave oven works is an amazing feat of engineering. Here is what happens inside it. A device called a magnetron generates microwaves that not only hit the food inside the oven, but go right through it. In a regular convection oven, hot air only hits the surface of food and heats it up. The hot layer then slowly heats up the inner layers via convection. In a microwave oven, the waves pierce right through. Remember the magical properties of water from Chapter 1? Water molecules are electrically charged, with the hydrogen ends being positively charged and the oxygen end being negatively charged.

All electromagnetic waves are essentially electric and magnetic fields changing rapidly at a particular rate. A radio wave does this at a leisurely 100 million times every second, which is why your favourite radio channel is at 104.8 MHz, while the X-ray your lab technician beams at you to see if you have a fracture changes its electric and magnetic fields at a slightly more energetic pace of a million billion (that's 1 followed by 16 zeros) times a second. A microwave does this in the range of 300 million to 300 billion times, which is still less than visible light, which is what our eyes use to see things. The thing to consider here is that you are packing all that radiation in a small, confined space. If you actually put a 2000 W halogen light inside a small box, it would cook food too, but it would waste a lot of energy heating up the box itself. We use microwaves because of a very specific property.

When microwaves at a specific frequency are beamed at things that contain water, they do something interesting. The water molecule, in the presence of the microwaves' rapidly changing electromagnetic field, flips back and forth as its positively charged hydrogen end and negatively charged oxygen end try and align with the direction of the field. And what happens when molecules keep flipping and moving around is that they gain heat energy. This is how a microwave heats up a glass of water in under 10 seconds, while your stove seems to take forever. The more observant readers are probably wondering that if something as low-energy as a microwave can heat up water inside foods, why can't visible light? In simpler terms, why doesn't water spontaneously heat up when kept outside, given the sheer amount of visible light radiation that surrounds us? Great question! To answer this, we need to understand a little bit of quantum physics. One of the most mind-bending realizations in theoretical physics in the twentieth century was this idea that, at the atomic level, energy is not continuous but works in well-defined spurts called quanta. Without getting into the gory details, the remarkable ingenuity of electrical engineers meant that they realized that water molecules could flip back and forth by absorbing energy from the microwaves, but only if the microwaves had a specific amount of energy, and not more or less. It turns out that the energy levels of visible light do not fit into the cosmically approved levels water molecules require to heat up. By the way, higher energy waves, such as X-rays and gamma rays, will heat up water, a phenomenon people who watched the HBO series, *Chernobyl*, will be familiar with.

So, enough physics. Let's get back to the kitchen. Microwave ovens work by heating the water inside food. Since most food contains a lot of water, it's a tremendously useful device. You can boil potatoes in it by placing them in a vessel with a microwave-safe lid (read: lids with small holes to let super-hot steam out) and a tiny bit of water, which accelerates cooking as it heats up and transfers this heat to the potato via convection, in addition to the microwave radiation. You can also cook rice and dal the same way, but you will need slightly more water because rice and dal by themselves have very little moisture. You can also steam idlis in a microwave oven because batters are mostly water, but you will need non-metallic, microwave-safe idli trays.

Here's a simple way to make an entire side dish in the microwave, using all the science principles we have learnt so far. Take a microwave-safe vessel, add some coconut milk (or yoghurt emulsified with some starch), some extra water to thin it out, a can of cooked chickpeas (discard the gooey water it swims in though), salt, turmeric, chilli powder, roasted cumin powder, onion and garlic powder, and a pinch of garam masala. A pinch of sugar will balance all the tastes. Now microwave this at the highest energy setting for 3 minutes. If you are using coconut milk, squeeze in a bit of lime juice at the end. You will get a surprisingly delicious chana sabzi for the amount of effort put in. The coconut milk, or yoghurt, is rich in fats. In 3 minutes, all the spice flavours will get dissolved in the fats and that's pretty much what you need. You can finish this off with a tadka and, voila, you have a 3-minute weeknight side dish!

You can improvise by adding steamed vegetables as well. Since 3 minutes won't be enough to cook many vegetables, it's better to steam them ahead of time and then add them to the dish. If you are wondering why you can't just add oil, spices, water and other ingredients and let it cook in the microwave, it's because oil is not heated up by microwaves, and it does not mix with water. Coconut milk/yoghurt are fat–water emulsions, and thus the water molecules heated up by the microwaves, will heat up the nearby fat molecules.

Here's a quick microwave cheat sheet:

1. You can cook any ingredient that has water in it. Most vegetables will cook in 3–4 minutes at high power (high power on your microwave is most likely the default setting, but do check the manual to learn how to change the power setting). You can coat them with a little bit of oil, and spices, to infuse some flavours too.

2. You can melt butter to perfect, spreadable consistency by reducing the power to half. At the default high-power setting, the water in the butter will turn to steam and break the

emulsion, leaving you with translucent butter that is not very spreadable.

3. When you try and roast papad in a microwave, you will find that it cooks unevenly. There's a nice physics explanation for that. The frequency that microwave ovens operate at is 2450 MHz. This means that the electromagnetic waves oscillate around two billion times in a second. And since microwaves, like all forms of radiation, travel at the speed of light, you can calculate the distance between two peaks of the oscillating wave by dividing the speed of light, which is a constant everywhere in the universe, by the frequency. That gives us a wavelength of about 12 cm. Since this is the distance between two peaks, it means that there should be two points where the field will be at 0.

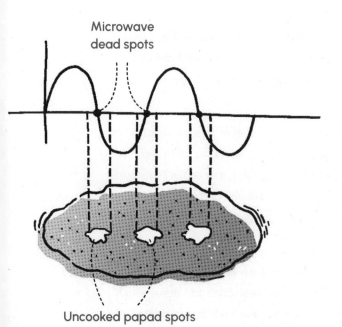

Microwave dead spots

Uncooked papad spots

As the microwave turntable spins, it reduces the chances of static dead spots

These are dead spots in every microwave oven, meaning anything placed there will not cook. Most microwaves solve this problem by having a rotating turntable, which ensures that no part of the food stays in a dead spot for the entire duration. But a papad is a large, flat circle, and some points will end up in these dead spots more often than not.

So, now that you know the physics, you can use it to your advantage. You can roll the raw papad into a glass tumbler and put it in the microwave. This way, every spot on the papad will rotate with the turntable and get even coverage. Now you know why restaurants in India tend to give you complimentary papad that looks tubular. They are microwaving it in a drinking glass which is the only way to ensure that it's evenly cooked.

4. You can also collect the malai (cream) that floats on top of milk every day. Once you have enough, just put it in a tall microwave-safe vessel (so that it does not overflow) and microwave it for 7–8 minutes. You will get ghee once you filter the brown milk protein.

5. If you crave chickpeas one day but forgot to soak them, worry not. Simply microwave the chana in water for 20 minutes (at a low power setting) and then let it sit for another 20 minutes in the same hot water. You will have chana that is as good as the one soaked for eight hours. Pressure-cook it and use it as you please.

6. You can also dry fresh herbs and turn them into powder in the microwave oven. Do this for curry leaves and mint, which usually have a very short shelf life, even when refrigerated.

7. Let's say you crave some rice late at night (who doesn't!). But who wants to cook a single portion in the pressure cooker and then have to clean all the vessels? What if I tell you that you can make instant pulao in the microwave just by adding some ghee, whole spices, washed rice, salt and water. Simply let it microwave for 10 minutes at high power, and then 10–15 more minutes on a low setting, and it will be done. If you remember the method to cook rice perfectly from Chapter 1, the first step is about gelatinizing starches, which happens at about 80°C. The second step is about lowering the temperature and letting the gelatinized starches absorb water till they become nice and soft. That's exactly what we are doing here too. Once done, squeeze some lime juice, because acids improve everything, and if you want to take it to the next level, sprinkle some MSG or mushroom powder for the umami hit. A caveat: don't try to feed your entire family. Remember how microwaves have 12 cm wavelengths and create dead spots? Using a large amount of rice risks uncooked or undercooked portions being left in some places, so this method is ideal only for a single serving.

Dehydrators

A common culinary tradition that urban India seems to have largely given up on, mostly due to the lack of open terrace access, is dehydrating food. Most foods concentrate flavour once they lose water. Anyone who has enjoyed the concentrated umami explosion when eating sun-dried tomatoes will attest to that. Likewise, yoghurt-soaked, sun-dried chillies and sun-dried citrus fruits make for fantastic condiments and pickles, while ground and sun-dried unripe mangoes become amchoor. The good news is that you can dehydrate foods at home, even if you don't have access to a terrace.

A dehydrator is not very expensive when you consider that you can dehydrate a ton of things, which would otherwise go waste, and turn them into long-lasting delicious snacks (dried mangoes or grapes, for example). You can also create your own spice powders like onion, garlic and green chilli powders. You can use the papery skins and outer layers of the onion that you normally discard, throw them into a dehydrator and turn them into onion powder, which can then turn your snacks into consumer-grade flavour bombs.

If you don't have a dehydrator, a regular convection oven at its lowest temperature setting will work too, but dehydration is a time-consuming process, so don't forget to factor that in.

Electronic Pressure Cookers

If you remember the chapter on pressure-cooking, the traditional way to do it is to throw all ingredients into the vessel, add just enough water and seal it tight so that no air escapes. Now, as the water boils, it turns into steam that has nowhere to go. Like unemployed youth getting into the drug trade, this steam then hangs around and prevents other water molecules from becoming steam by increasing the air pressure inside the cooker. This increases the boiling point of the water inside to 121°C, thus allowing food to cook faster. Simple. Except, this device needs careful attention. You need to ensure that once peak pressure is reached and the whistle blows, you reduce the heat to ensure the pressure stays reasonably constant. You also need to remember to turn it off before all the water boils off and your food starts burning, and you need to wait patiently for the steam pressure to come down before opening the lid. If not, you run the risk of getting a high-temperature steam facial.

What the Instant Pot, or to use the more general term, an electronic pressure cooker, does is that it turns pressure-cooking into a fill-it-shut-it-and-forget-it experience, which significantly reduces the anxiety that inexperienced cooks tend to have when it comes to pressure-cooking—the subliminal fear that you have a potential bomb on your stove. How it works is by using sensors that keep a close watch on two things: temperature and pressure. Once you put your ingredients in and shut the lid, you are presented with several simple options that take the complexity out of pressure-cooking, not that it was terribly complex to begin with. There is a neat bit of engineering involved, which readers of this book will easily understand.

Remember the boiling point of water? Most cooking happens under 100°C because once the water boils away, you don't have a cooking medium for heat to transfer evenly to your food. Yes, the pan can transfer heat, but it transfers a lot of it. This is because metals, as we learnt, are good conductors of heat and that can quickly burn the outside of your food before the inside cooks. So, the first engineering breakthrough to get past the 100°C limit was the traditional pressure cooker, which let you cook at 121°C by increasing the pressure inside the vessel and thus increasing the boiling point of water. The big problem with pressure cookers is the anxiety of whether or not you have added enough water. Add too much water and you get mush, put too little and your food will burn because all the water will turn to steam and your food will be in direct contact with hot metal. The electronic pressure cooker solves this problem by using a temperature sensor. The moment the temperature crosses 121°C, we know that the water has probably turned to steam, so it's direct transfer of heat from the metal to food. This is when the sensor shuts the heat off. This allows you to use the least amount of water to pressure-cook food, and that will always result in better texture and mouthfeel. Remember, a carrot is 88 per cent water, so you don't need too much water to pressure-cook it.

What it also does is help you parallel process in the kitchen. Once you've shut the lid, you don't need to worry about when the cooking process will end. Once you've set the timer, it will cook at peak pressure for that long, reduce heat and slowly release steam, and then maintain the dish at a warm-enough temperature for you to serve.

Another useful feature that several electronic pressure cookers have is the ability to maintain a constant temperature, which allows you to make yoghurt with consistent thickness and sourness. Several models come with a 45–46°C 'yoghurt' setting that lets you put in the milk mixed with the starter culture and notifies you when the yoghurt is done.

Modernist Ingredients

Now, if you are comfortable with sodium chloride, sodium bicarbonate and monosodium glutamate in the kitchen, I'd like to introduce you to a few more sodium salts that will make you popular with guests, particularly those who come over to watch IPL matches. The first of these is sodium citrate, the salt of the acid found in citrus fruits. Predictably, it is sour-tasting and, because it's a sodium salt, salty. In fact, it's commonly called sour salt, a common preservative used by the consumer food industry. Read the label on literally any product and you will find sodium citrate listed. It riles me up when people say silly things like, 'I don't like to eat food with all those chemical preservatives.' My instinctive response is, 'Would you rather eat food that goes bad quickly?' But at the same time, I agree that store-bought ginger–garlic paste tastes terrible because the sodium citrate lends a sour taste to it that does not go well with the complex flavours of ginger and garlic.

Irrational worries aside, sodium citrate in your kitchen can play a different role. Not that of a preservative, but that of an emulsifier. When you are watching IPL matches, you are probably gorging on crispy snacks that are best had with sauces. If you are among those who pour out some ketchup in cups, I'll give you 4/10 in hospitality. If you make a fresh green chutney with coriander, mint and chillies, I'll give you 8/10. For that perfect 10/10, you must, in addition to the chutney, make a creamy, desi-flavoured cheese dip. You can, of course, buy these at a store. They are expensive, and after that IPL party will sit in your fridge, contemplate on the meaninglessness of life, lose moisture and turn brown. But making cheese dips presents a problem. The best-tasting cheese won't melt evenly. The fat will tend to separate from the proteins, resulting in a hard glob of protein sitting in a pool of melted fat. This is exactly where sodium citrate works its magic. A tiny pinch of sodium citrate will let you melt any fancy block of grated cheese, giving it a delectable creamy texture. Now, add some spices and flavouring to this, and you will have a fantastic desi cheese dip to serve, one that your guests will lick clean.

Because of the title of the chapter, I'm also obligated to say that you can take this to the next level and make your own desi-flavoured cheese slices, which again, are rather expensive if you choose to buy them. All you need to do is bring some beer to a simmer, add a pinch of sodium citrate and some grated good-quality cheese. Let it thicken into a creamy sauce. Add finely chopped chillies and chaat masala (or idli podi) and pour it out on a baking tray, ensuring a thin, even layer. Let it cool in the fridge for a few hours and then slice it into squares. Voila! Your chilli–chaat cheese slices are ready!

A common problem when making large batches of any kind of green chutney is the discolouration because of a combination of factors, one being the chlorophyll molecule losing its magnesium atom and the other plain old oxidation. This discolouration will happen even in the refrigerator, so here are some tricks to avoid this. We already learned about one of them (blanching greens for 30 seconds and then shocking them in an ice bath to deactivate the enzyme responsible for stealing the magnesium from chlorophyll). But even this will keep your chutney green for a few hours at best. Enter our second modernist ingredient, sodium bisulphite. A really tiny pinch of this will keep your greens looking bright for a really long time. This molecule prevents most enzymatic browning (caused by polyphenol oxidase) and non-enzymatic browning (caused by good old oxygen). Added bonus: It is also a preservative that will prevent fridge-friendly fungi and bacteria from dipping into your chutneys.

The third ingredient I urge you to keep around in your kitchen is xanthan gum. Despite sounding like a sticky alien that hails from a planet in a faraway galaxy, it is a polysaccharide (long chain of sugar molecules) produced by the fermentation of simple sugars by a bacteria named Xanthomonas campestris. This magical substance, even a tiny amount, can thicken any gravy, batter or dough. You might wonder why not just use corn starch, rice flour or maida, but that's like asking why use an Uzi submachine gun when a blunt knife is available. If you have ever tried making rotis using millet-based (or any non-gluten) flours, you know how painfully difficult it is. Try adding a pinch of xanthan gum and you will be able to make jowar and bajra rotis in shapes that don't resemble Yugoslavia at the peak of the Balkan crisis.

Another modernist ingredient is soy lecithin. Just take a look at any creamy-textured food product. It will probably have soy lecithin as an emulsifier. To understand what it does, let's understand what emulsification is. Typically, fats and water do not mix. But when you have a phospholipid, which is a fat molecule with a phosphate group attached to the glycerol backbone, in addition to the long fatty acid chains (recall Chapter 1), it allows the molecule to do something interesting—the phosphate side of the molecule binds with water, while the fatty acids like to hang out with their oil friends. So, when you introduce an emulsifier to a mixture of water and fat, and mechanically agitate it, it will turn into a creamy emulsion, like mayonnaise. Lecithin is a phospholipid found naturally in egg yolks and soy lecithin is simply lecithin produced from soybeans. You can use it to make super-stable salad dressings and also increase the shelf life of anything you bake in an oven.

By the way, phospholipids are how our bodies digest and transport fats. We are mostly water, and fats don't mix with water, so the phospholipids in our intestines emulsify fats (literally like mayonnaise) to be able to transport them easily.

Smoking

Since the time the caveman figured out that meat hung to dry in a smoky environment improved its taste and increased its shelf life, smoking has come a long way. It is now an advanced science, particularly in the case of southern United States, which has strict rules about what wood to use, what cuts of meat to use and for how long. In this part of the world, smoking is a natural outcome of cooking using firewood, which is common in rural parts of India, and cooking food in a tandoor oven, which uses charcoal. There is a jugaad you can employ to impart a tandoor-style smoky flavour at home. You'll need to heat up a piece of charcoal for about 5 minutes, then place it in a small metal cup, drop a teaspoon of ghee into it and place this in your dish, sealing the lid for 2–3 minutes.

The smoky flavours from tandoor-style smoking and open cooking on a wood fire are entirely different. The charcoal smoke flavour does not come from the charcoal; it comes from the fats in your food (twenty aromatic compounds), which then waft up with the smoke and attach themselves to the food. On the other hand, wood is made up of cellulose and hemicellulose, with lignin acting as a glue of sorts. Cellulose and hemicellulose are carbohydrates, essentially made up of long chains of simpler sugar molecules. Now, recall what happens around 160°C. Sugars caramelize, and that's exactly what happens here too. The cellulose breaks down into several aromatic compounds that releases the characteristic sweet, floral and fruity aromas associated with the caramelization of sugar. Lignin, on the other hand, produces several distinctive aromas, such as vanillin (the primary flavour of vanilla) and other clove-like flavours. So, when certain kinds of hardwood are burnt at relatively low temperatures, they produce fantastic aroma molecules that lend complex flavours to the food being cooked over them. If the wood is burnt at too high a temperature, it produces acrid, carcinogenic substances that you should keep your food away from.

Around a hundred years ago, some entrepreneurs figured out that they could slowly smoke wood, distil this smoke and get it dissolved in water, essentially creating wood smoke essence. This substance is called wood vinegar, although you should call it pyroligneous acid to appear cooler and better informed. Adding this liquid to cooked food will impart the same smoky flavours without the mess of burning wood and dealing with asphyxiating smoke. You can buy this liquid smoke (which is pyroligneous acid mixed with some soy sauce for the umami and salt hit) online and add a tiny bit to any cooked dish.

Tip: Don't go overboard because it then becomes easy to detect that you've taken the shortcut. We don't want that, do we?

Sous Vide

If you recall Chapter 1, we learnt that the following things happen to food at the following temperatures:

1. 40–50°C: Proteins in fish and meat begin to denature.

2. 62°C: Eggs begin to set.

3. 68°C: Collagen in the connective tissues of meat denatures.

4. 70°C: Vegetable starches break down.

5. 110–154°C: Maillard reaction becomes noticeable.

6. 180°C: Sugar begins to caramelize visibly.

A couple of things become evident. Compared to plant products, meat is very sensitive to heat. But given the tremendous richness of spices that Indian cuisine tends to use, for the most part, home cooking tends to not bother too much about the texture of meat. The tendency is to play it safe and cook it till it's hard. The second thing is that every ingredient has a different ideal temperature, at which it is just the right combination of doneness and texture. Typically, we adjust for this by sequencing how we add the ingredients to our cooking vessel. The things that cook quickly are added later, but there are other problems. Ingredients tend to cook from the outside in, and depending on how much heat you apply, you can end up not cooking the interiors enough, or overcooking the exterior.

Since this chapter is about taking your cooking skills to the next level, there is a way to assemble a dish where every ingredient is cooked perfectly, and like electronic pressure cooking, it's pretty forgiving in terms of time. This technique is called sous vide, which means 'under vacuum' in French. The vacuum bit is more or less optional, as long as you seal the ziplock bags that you use to cook the food by pushing out as much of the air as possible. If you are wondering, 'Wait, are we going to be cooking food in ziplock bags now?' the answer is, 'Yes, it works. Just hear me out.' Let's start with the basic principles of physics behind sous vide cooking.

1. Heat flows from an object at a higher temperature to an object at a lower temperature.

2. Once both objects reach the same temperature, heat stops flowing. This is called thermal equilibrium.

In sous vide cooking, we use this idea to heat up a water bath to a precise temperature. This is done using a device that, frankly speaking, is a fancier version of the immersion heater Indians use in the bathroom. Okay, I can see how some of you are going, 'Ziplock bags and bathroom immersion heaters? Ashok, this does not sound palatable.' But trust me on this.

The actual device is a precise immersion heater that, in addition to maintaining the temperature of water, also keeps circulating heat so that anything cooked in the water bath has heat transferred to it evenly. The food itself is not dropped directly into the water, but is put in tightly sealed, food-safe ziplock bags. This is because we only want the heat from the water, not the water itself.

When we cook anything in a normal vessel containing water, the water closer to the heat source will be hotter. This is why the sous vide device constantly circulates the water to maintain an even rate of heat transfer. Once your food has reached the temperature you want it to be cooked at, say, a chicken breast at 65°C, no further cooking will happen even if you keep the device on, thanks to the principle of thermal equilibrium.

It does take time to get to equilibrium though, so this is not a technique for a last-minute dish. The advantage is that you can leave things in the sous vide setup and do other things simultaneously, knowing for sure that it will not be overcooked.

What is sous-vide?

Slow cooking technology for the perfect meat

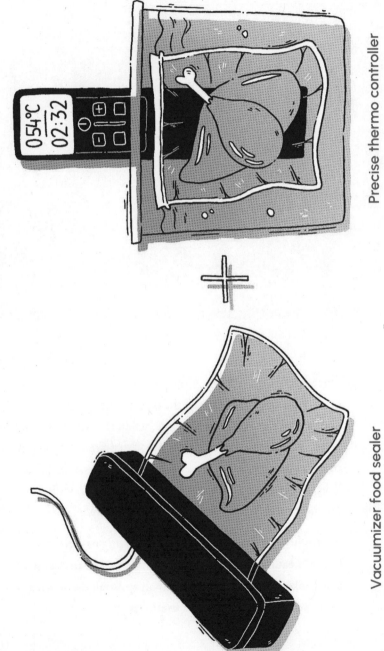

Precise thermo controller

Vacuumizer food sealer

185

Item	Cut	Temperature °C	Time (mins)
Chicken	Breast Thigh Legs	64 69 69	90 120 150
Mutton	Tender cuts Tough cuts	70 70	150 480
Pork	Tender cuts Tough cuts	60 75	120 480
Beef	Tender cuts Tough cuts	55 75	120 480
Vegetables		85	20-60

Once you are done, there is one last critical step. Remember how while most cooking happens under the boiling point of water (100°C), the magic happens in the 110–154°C range, when the Maillard reaction happens. When you cook chicken in a sous vide cooker, it will come out looking pale and unappetizing, but it will be perfectly cooked through and juicy. What is needed is a short visit to a really hot pan with some fat, so that the meat can sear into a gorgeous golden brown.

Here is how you can make the most amazing modernist chicken tikka with breast meat, which is the most frustrating cut to deal with because it overcooks way too quickly. We shall apply all the science lessons we have learnt so far, turning it into a useful recap.

1. Brine the chicken pieces in a solution of salt, dry spices and water. A 5–10 per cent salt solution will work. Try various levels in between to arrive at what works for you. You will need to do this for two hours for every 1 kg of chicken. Once you are done, wash it in plain water.

2. Now, marinate the brined chicken in ginger and garlic paste, a little bit of salt, garam masala, chilli powder, turmeric, oil and yoghurt in a bowl for about an hour.

3. Place a small cup in the bowl, drop in a piece of charcoal heated for at least five minutes and pour a teaspoon of ghee on it to start the smoking. Seal the bowl with a tight-fitting lid to keep the smoke inside. I'd say smoke it for at least 2 minutes. You can do it for longer if you like a stronger flavour.

4. Transfer the marinade to a food-safe ziplock bag and drop it into a sous vide bath set to 67°C for about 90 minutes. If you don't have a sous vide device, or a water bath, no worries. Here's the perfectly serviceable jugaad: Take the largest stock pot you have and fill it with water, up to three-fourths of its depth. Let it simmer and then turn the heat down to sim. Now insert a temperature probe and wait for the water to reach 70°C. At this point, lower your ziplock bag into it in such a way that all the chicken is under the water. Secure the bag against the wall of the vessel with a metal clip. Stir the water once in a while and keep a watch on the temperature. If it goes below 65°C, use the lowest heat setting to get it back up to 70°C. Do this for 90 minutes. Yes, it sounds painful, but this is what you need to do to get the most succulent chicken breast possible. If you really like the output, you can always buy a cheap sous vide device online.

5. Take the cooked chicken out and grill it in a pan with some oil, at a high temperature, till it browns beautifully.

6. You can now serve it as a dry dish, after sprinkling some chaat masala on it for the mild, pungent hit of the hydrogen sulphide and a little lime juice for the acid hit. Or, you can drop it into a makhani gravy (Chapter 7) to turn it into chicken tikka makhani.

7 Burn the Recipe

Knowledge is knowing tomato is a fruit, wisdom is not putting it in a fruit salad, and philosophy is wondering if ketchup is a smoothie.

—Anonymous

Picture yourself as a willing student of music. An eager-eyed, excited student willing to do what it takes to learn this art, and then the teacher tells you, 'Start with a 120 bpm tempo and play the C, Am (in first inversion), F (in second inversion) and G (in first inversion) chords in ostinato style on the piano.' And you go, 'But I only have a guitar', to which the teacher says, 'No. This is how the music is written and it shall be performed in exactly that way.' So, you go buy the cheapest Casio keyboard and then look up instructions for how to play the Am chord. You first learn the basic version, then the first inversion and finally how to play it in ostinato style, but the teacher goes, 'No. The chord needs to span five octaves, so please get a proper grand piano.'

That teacher is a recipe. And recipes are a terrible way to approach cooking. Okay, I'm being unnecessarily harsh here, so let me dial it back a bit. Good recipes that use simple ingredients are a decent way to get started on your journey to becoming a good cook. In general, recipes are limiting from a culinary education standpoint because it's like trying to learn a craft by only looking at the output, with no knowledge of why what you do has the effect it does.

This problem is particularly exacerbated in a culture as diverse and rich as the subcontinent, where every single recipe lays claim to being authentic. If you ask me, I'd say this is a silly idea when it comes to food. Every recipe will use slightly different methods and insist that there is this one specific step that makes all the difference (it usually doesn't). Also, a majority of recipes is largely untested in terms of weights and proportions.

In an ideal world, those publishing recipes will A/B test multiple methods for readying every ingredient, combinations of spices, flavour extraction methods and proportions of ingredients and spices. This is what *America's Test Kitchen* or *Serious Eats* does. They sometimes test hundreds of recipes for a single dish, like a grilled cheese sandwich, and put out a reasonably authoritative opinion on what kinds of cheese, and what methods (butter or mayonnaise), will make the perfect sandwich. They also regularly A/B test common cooking myths (like 'searing the meat seals the juices'; it does not). There is no *Indian Test Kitchen* as of today, but it's also significantly harder to pull off. A majority of Indian dishes tend to involve a large number of ingredients and multiple preparatory steps, each of which impact the final flavour. As a result, there are many paths one can take to get to a palatable product. There is no perfect recipe for any dish. I do hope though that there is a test kitchen for the more minimalist, super-sensitive-to-technique dishes such as idli, chapatti, dosa, pulao, etc.

What this chapter will attempt to do is to help you put in place a practical, minimalist test kitchen protocol for your home, one that helps you become a more scientific-minded cook and liberates you from the chaos of recipes. To do that, we need to understand some simple principles.

A/B testing is a user experience research strategy that involves a randomized experiment with two variants, A and B. In the context of the kitchen, A could be a dal with amchoor as the acid, while B could be lime juice. A could also be using minced garlic while B could be garlic paste. The idea here is to use every night's cooking as a vehicle for multiple As and Bs, while you document the results. In some cases, if you are really up for it, you could split your dal into two and use amchoor in one and lime in the other, and see which one you, or your family and friends, prefer. But you cannot always do the full A/B test in a single meal. This is because minced garlic, as opposed to a paste, goes in right at the start of the dish, so in these cases, do A one day of the week and B on another day. It then becomes useful to document what you are doing every single day, so that you can compare notes. I know this seems like too much work, but you only need to do this once for each cooking method and spice combination.

The second idea to understand here is the concept of a metamodel. The recipe for paneer butter masala is a model. The generalized method for a (makhani/Punjabi/Chettinadu/Malabar-style) gravy dish featuring either vegetables or proteins is a metamodel. This chapter is all about metamodels and generalized algorithms for the most common patterns of Indian dishes. A big J.B.S. Haldane paraphrasing caveat: The subcontinent is not only more diverse than we imagine, it is more diverse than we can imagine.

What we are attempting to do is not a grand unified theory of Indian cooking; even Einstein failed to do that for physics. What we shall explore is a more modest, and hopefully more practical, special theory of Indian cooking (conditions apply).

Special Theory of Indian Cooking (Conditions Apply)

Let us first get the conditions out of the way. This generalized set of algorithms will not cover every single culinary tradition in India. It will, for the most part, restrict itself to urban, middle class India, the kinds of ingredients that are likely to be available and cooking techniques that are practical in a smallish apartment. So, no tandoors, no wood fire and no nomadic horseman-style dum cooking by burying meat and rice into the ground with coal embers.

The second condition is that culinary traditions in India not only vary across state and linguistic boundaries, but also by caste and community, which is why the examples here will largely be restricted to the kinds of dishes available in run-of-the-mill restaurants. So forgive me if I have missed out Cudappah cuisine while including Hyderabadi. The intent here is to arm you with a way of thinking that will help you make a specific dish from, say, Odisha with confidence. The algorithms themselves may not cover every single sub-cuisine in the country. If this chapter ignores your community and state's cuisine, it's not deliberate. The examples are for representative purposes only. You can instantiate a version of this for your cuisine rather easily.

The third and final caveat is that we shall keep aside that universe within a universe of starters, snacks and tiffin items, because trying to cram that in will be the equivalent of boiling the Indian Ocean. Instead, we shall stick to gravies, rice dishes, breads, chutneys/raitas and salads. These algorithms will give you a wide-enough repertoire to start with. The rest of the journey, as always, is up to you. Treat this like high-school science education. University is on you.

So, the special theory of Indian cooking starts with the all-important question: What do you want to cook? Depending on your answer, you can opt for the following paths:

1. The Indian gravy algorithm: This will present a generalized algorithm and metamodel for preparing vegetables, legumes, meat or eggs in a sauce-like gravy that is flavoured in a specific regional style, like Malabar, Punjabi or Bengali.

2. The rice dish algorithm: A generalized method for preparing steamed rice, flavoured rice, khichdi, pulao and rice for biryani. There are numerous other ways of cooking rice in the subcontinent, but these five are the most utilitarian.

3. The Indian bread algorithm: Standardized and consistent methods for preparing doughs for unleavened breads (chapatti and paratha), leavened breads (naan and kulcha) and non-gluten-based breads (bajra or jowar roti). We will stop at the dough stage because rolling and baking/tawa operations are better learnt by watching an experienced hand. You can't learn it from a book.

4. The chutney and raita generator: A metamodel for generating your own chutney and raita recipes from whatever ingredients you have available.

5. The salad generator: A metamodel for generating your own salad recipes by hitting the right balance of greens, crunch, protein, acid and flavouring.

The second question to ask is: How do you want to make this dish?

1. I'd like to see what's in my fridge and pantry and make the best of it.

2. It's my wife's birthday and she is from Panjim, so I am looking to make a dish that evokes a specific regional cuisine, say pork vindaloo.

Once you have the answer to this, you need to execute Step 0, which is prepping the ingredients. After all consistency and productivity require you to approach home cooking the way restaurants do it. Also, prepping is not just cleaning and chopping, it includes a whole range of activities from brining to marinating to steaming and sautéing, all of which will make you a better home cook. In fact, a lot of prep work is actually cooking for the most part.

Prepping for Productivity and Maximizing Flavour

Regardless of the kind of dish you want to make, prepping will help you execute a plan. The French call this the mise en place, literally 'putting in place', and professional kitchens in a restaurant do this every single day, depending on the items listed in the menu. When you order butter chicken, here is what happens in a professional kitchen. One chap retrieves some finished tandoori chicken spinning around in a rotisserie (that keeps it warm) and chops it up into small boneless pieces. Alternatively, he might pick a prepped kebab stick with marinated boneless chicken and stick it into the tandoor for a minute or so. The 450°C oven will cook the chicken rapidly while ensuring it stays juicy. He then keeps the just-cooked pieces ready for your dish. Another chap retrieves a base makhani gravy that was prepared earlier during the day from a big stock pot. In a small fry pan, he heats some butter and cooks the gravy a bit. He then adds a few spice powders, dunks in the prepped chicken pieces, some fresh cream, some fenugreek, a glob of butter and chopped coriander for garnish.

We aren't aiming for that level of industrialization, but what we are trying to do is pick up the habits that help us make delicious food efficiently, while ignoring the things that make restaurant food unhealthy. As I said at the start of the book, we are trying to bring in some craft to what people tend to consider an art. And craft requires planning and preparation.

A general rule to keep in mind is that things that cook at different rates are better off prepped and cooked separately, before they are ultimately brought together to finish a dish.

Vegetables

Since onions and tomatoes tend to be rather universal and play a hybrid role as a flavouring agent/acid, we shall treat them separately. Don't forget to wash your vegetables!

As far as onions go, they can be:

1. Cubed (for mild flavour, best added later in the dish)

2. Half-moon sliced (for mild flavour)

3. Minced (for medium flavour)

4. Pureed (for high flavour. Add this earlier in the dish and cook it for longer)

Some handy tips:

1. If you want to reduce the lachrymatory torture because of the syn-Propanethial-S-oxide generated each time you mechanically damage an onion, cut it right under a ceiling fan, or use a small USB fan to blow away the irritant. However, the fan will simply blow it all over, leaving the other occupants of your house with some mild to moderate eye irritation. Depending on the general political philosophy at play in the household, this might be acceptable as an instance of 'let's share both the pain and joy' socialism, or unacceptable from the standpoint of 'You suffer the pain, keep me out of it, but give me the finished product to enjoy' exploitative capitalism. You could also chop onions under water and avoid this political dilemma altogether.

2. In most situations, you can replace shallots with onions. Shallots have a sweeter flavour profile that works perfectly with hot and spicy south Indian dishes, so if you can get them, nothing like it. Please ask misbehaving adult children to peel shallots as penance.

When it comes to tomatoes, here are your options:

1. Chopped (for gravy dishes with a bit more texture or for drier dishes).

2. Pureed (for smooth gravy dishes).

3. To take things to the next level, add concentrated tomato paste.

4. If you don't have tomato paste, use a sachet of ketchup.

5. In dishes where the tomato is the star of the show, consider adding some dehydrated tomato powder too.

Some handy tips:

1. Recipes that call for removing the seeds and pulp should be burnt at the stake. That's where all the flavour is. Of course, there might be some situations that call for removing the skin, although you can always puree and run it through a sieve to remove the fibrous bits. The outcome will largely be the same.

Vegetables can be:

1. Peeled if they have a thick, inedible skin. In general, we tend to discard more skin than necessary, and almost all vegetables have a ton of flavour just beneath the skin. You can use a general rule of thumb that vegetable or fruit flavour is a gradient that goes down as you travel from just under the skin to the core.

2. Cut:

    a) Chopped to the size you want in your dish.

    b) Minced, if it is going to be cooked for a short time.
       Remember, the more the surface area, the faster it will cook.

Once you have done that, you can also:

1. Use the chopped vegetables as is.

2. Blanch: This is best for green leafy vegetables. You can also puree them, say, for palak dishes.

3. Steam: Generally better than boiling vegetables.

4. Sauté in fat: To brown them and add more flavour, thanks to the Maillard reaction.

5. Bake: Coat them with oil and put them in an oven at 180°C till they are evenly browned. This is more time consuming but requires less fat. If you have an air fryer, you can do this faster than in a conventional oven.

6. Deep-fry: Reserve this for special occasions and obsessive attempts to 'get that restaurant flavour'. When you order a bhindi masala in a restaurant, the bhindi is chopped and deep-fried before being added to the gravy. That is why it tastes so good.

Some handy tips:

1. Don't waste your time marinating vegetables. They usually don't have enough protein for the acid to tenderize, and they absorb salt pretty quickly in a dish anyway.

2. It's a good habit to store chopped vegetables in a bowl of water. While some vegetables do not oxidize quickly, many do. Research on habit-forming tells us that it's always better to apply a rule to everything so that the habit is easy to institutionalize. If you are chopping vegetables well ahead of the cooking time, squeeze some lime juice into the water. Vitamin C (ascorbic acid) is an excellent antioxidant.

3. Don't throw away peels. Collect them and store them in the fridge to make vegetable stock. Once you are done using them for stock, you can compost them.

4. Be careful with green/leafy/delicate vegetables. Tearing the leaves by hand will cause damage in between cell walls and help retain crunchiness and structure. Using a knife will likely slice through cells and activate enzymatic defence mechanisms that cause them to wilt and brown.

Fried Snacks

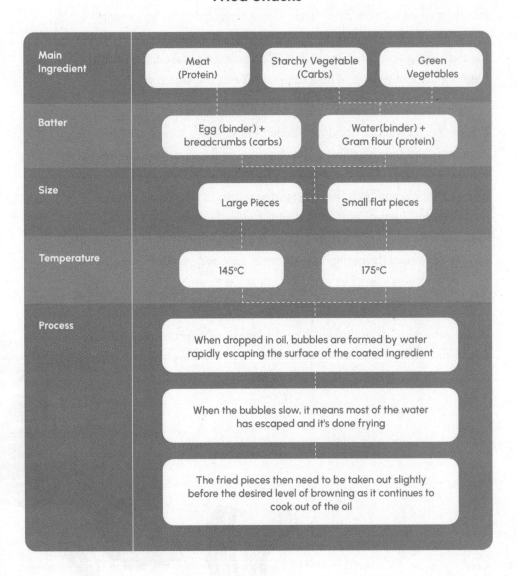

Main Ingredient	Meat (Protein)	Starchy Vegetable (Carbs)	Green Vegetables
Batter	Egg (binder) + breadcrumbs (carbs)	Water(binder) + Gram flour (protein)	
Size	Large Pieces	Small flat pieces	
Temperature	145°C	175°C	
Process	When dropped in oil, bubbles are formed by water rapidly escaping the surface of the coated ingredient		
	When the bubbles slow, it means most of the water has escaped and it's done frying		
	The fried pieces then need to be taken out slightly before the desired level of browning as it continues to cook out of the oil		

Meat and Seafood

As impressive a skill as cutting meat and cleaning seafood is, just get the butcher to do it for you. They tend to have better tools and know what they are doing. For the purpose of this book, I will assume that you have the cleaned cut you need for the dish. Elaborating on prepping meat will require a book of its own. There are several YouTube videos you can refer to if you really want to do this from scratch.

Prepping meat is about deciding whether to use it as is, to marinate it, or to brine and then marinate it. Just marination does very little, so my recommendation is to always brine as a bare minimum. Also, don't marinate for over an hour. Brining is what results in juicy, flavourful meat. You can always add flavour during the cooking process. Most cuts of meat, as we learnt in Chapter 1, cook at a low temperature. You can include additional flavouring in your brine, such as sugar and spices. In general, use spices similar to what you might use in a marinade. Only use them in smaller quantities because you don't need a lot here.

If you leave the meat in the brine for too long, it will get too salty. Keep this in mind and err on the side of caution here. You can also use a lower amount of salt and then choose to brine for longer.

Brining table for an 8 per cent brine

ITEM	CUT	TIME (MINS)
Chicken	Pieces Whole	60 180
Mutton	Pieces Large cuts	60 240
Pork	Pieces Large cuts	60 240
Beef	Pieces Large cuts	60 240
Fish		30

Legumes

Most lentils take longer to cook in an acidic gravy, which almost all gravies tend to be, so it's better to cook them separately. Pressure-cooking is the most convenient way to prep lentils.

You can refer to the table in Chapter 1 for soak and pressure-cook times.

1. Use a pinch of baking soda for harder/bigger legumes to reduce cook time.

2. Avoid adding salt before pressure-cooking to reduce cook times.

3. You can cook smaller lentils in the microwave after soaking them.

4. Add some fat when pressure-cooking lentils to prevent foaming.

Rice

Rice needs to be washed thoroughly till the water runs clear. This is because, in addition to the loose amylopectin starch that makes grains stick together when cooked, white rice tends to have other additives (like talc) that are used in the polishing process, none of which you want in your tummy. Once washed, you can either choose to soak it or use as is. A general tip is to soak the long grain varieties, used in pulao and biryani, for 20–30 minutes, so that a good amount of water absorption happens before the cooking process. This will result in a more evenly cooked final product. For making regular steamed rice, no soaking is required. However, in my experience, unless you are using a pressure cooker (which is perfectly fine given the convenience), go ahead and soak all varieties. This will reduce the cook time and minimize chances of the rice sticking to the bottom of the pot and getting scorched.

Eggs

You can choose to either drop eggs right into a gravy, or in a dry dish towards the end of the cooking process, or even prep them separately to ensure that you have the perfect texture you need. A quick recap of how to prep eggs is the best possible way to proceed. You can use eggs in any of these ways in your final dish (dry, gravy or, in the case of boiled eggs, even salads).

1. Boiled eggs: Bring water to a boil, drop the eggs into it, turn the heat off and close the lid. It takes 8 minutes for medium-boiled and 12 minutes for hard-boiled eggs to be ready. Adding a pinch of baking soda will reduce the chances of the egg sticking to its shell.

2. Scrambled eggs: Break the eggs and salt them for 15 minutes before cooking. Then, at low heat, cook the eggs in butter. Stir constantly. You can also add some cream or milk for fluffiness.. Then, at low heat, cook the eggs in butter. Stir constantly. You can also add some cream or milk for fluffiness.

3. Poached eggs: Bring water to around 95oC, just short of a full boil. Strain the egg of any loose egg white and drop it into the water for 5 minutes, till the white sets.

4. Omelette: Break the egg and salt it for 15 minutes before making an omelette. Add cubes of frozen butter when the omelette is cooking for a creamier texture.

The Rice Dish Algorithm

So, we are ready for our first algorithm. We shall begin with rice because rice dishes can be entire meals by themselves, especially if you are making a vegetarian/meat pulao, biryani or khichdi. If there is one category of dishes worth mastering the science of, it's the rice dishes.

What kind of rice dish are you planning to make?

1. Plain steamed rice.

2. Rice mixed with flavouring, like lemon rice, tamarind rice, capsicum rice, etc.

3. Rice mixed with lentils, such as pongal or khichdi.

4. Pulao, which is a rice dish where the flavouring and the rice are cooked in a single vessel in one shot. Since some of you will ask, 'But what about mutton pulao?', let me just say that

debates about culinary semantics are silly and this definition is meant to be a broad guide to help you think clearly in the kitchen. This is India, where there will always be exceptions with several dishes sitting on the fences of categories.

5.   Biryani, a rice dish where the flavouring and rice are partially cooked separately, after which they are layered and finally cooked. It is one of the greatest culinary inventions of the world. We have an entire chapter dedicated to this, so we will skip it for now.

Please note that when an algorithm calls for prepped ingredients, refer to the previous section.

Steamed Rice

Here is the algorithm for plain steamed rice:

1.   In a vessel, add prepped (washed and soaked for 20–30 minutes) rice, and water that is one knuckle of your index finger above the level of the rice. Bring it to a boil. Salt is optional. Keep in mind that a 1–1.5 per cent of salt by weight of final cooked rice is a good place to start. So, if you are a beginner cook breaking into a sweat about how much salt to add, weigh the water and rice, and add 1 per cent of salt. If that's not enough, add more the next time.

2.   Once the starches gelatinize and the water level reaches the top of the rice, reduce heat to low and close the lid. Wait for 10–15 minutes. Using a fork, gently check to see if there is any water at the bottom of the vessel. Do not fluff up the rice now, just do this surreptitiously in one corner. If there

is water left, close the lid and wait for another minute before checking again. Do this till there is no water left at the bottom.

3. Turn the heat off, keep the lid closed and let the rice sit for 10 minutes. This is crucial as it is at this point that the gelatinized amylose and amylopectin strike some backroom political deals and align to ensure that your grains are separate and fluffy.

4. Use a fork (never a spoon!) to fluff up the rice before serving.

If you are pressure-cooking, there's little to do once you put the prepped rice and water in. Depending on the amount and variety of rice, the altitude and your pressure cooker, you might need to tweak a few variables, such as the amount of water and cook time at peak pressure, a few times before getting it right. The open pot method is more forgiving as it allows you constant peeks at the rice, letting you adjust heat and time. In general, it's not a good idea to try and see what's going on inside a pressure cooker. Also, changing heat levels has little effect once peak pressure is achieved.

If you are cooking a very small amount of rice, just for yourself (and maybe one other person), microwave oven is an option too. Just 10 minutes at high and 10–15 minutes at low will get you perfectly serviceable steamed rice. Don't use this method for large amounts of rice though, as your microwave's dead spots will leave you with an unacceptable amount of uncooked rice.

Once you master this, you can start experimenting with flavours.

1. Use 50 per cent coconut milk + 50 per cent water.

2. Use soy sauce instead of salt for umami-flavoured rice.

3. Add flavouring to the water: Onion powder, garlic powder, lime zest (but never lime juice).

4. Use vegetable, meat or seafood stock to add more depth of flavour.

Flavoured Rice

The algorithm for flavoured rice is dead simple.

1. Make steamed rice using the method described above. Let it cool down.

2. Prepare a flavouring of your choice and mix it with the rice.

The flavouring here could be:

1. Tadka/tempering with whole spices.

2. Acid (such as lime juice).

3. Crunch (roasted peanuts, cashew nuts, lentils).

4. A full chutney or gravy (such as a tomato and onion chutney), or tamarind chutney.

Rice + Lentils

Typically, the texture of the rice in such a dish needs to be soft and well-mixed with the lentils. A pressure cooker is the most efficient way to do this. Here's how you can go about it:

1. Heat fat and whole spices.

2. Add prepped rice, lentils, stock (or water) and salt. With such a dish, you can be generous with the water because you want a mushy consistency.

3. Pressure cook for at least 15 minutes.

4. Once depressurized, open and add the finishing flavouring, tempering and herbs. Finishing flavouring could very well be a separate gravy, like in the case of bisi bele bhath.

Pulao

The algorithm for pulao is a fusion of the algorithms for flavoured rice and regular steamed rice. You might want to consider using a long-grain variety of rice, which has less amylopectin and whose grains stick less to each other.

1. Heat fat and add whole spices like cumin.

2. Add your choice of prepped vegetables. Use finely chopped vegetables so that they cook at the same time as the rice, while maintaining some texture and mouthfeel.

3. Add your choice of brined and marinated meat.

4. Add the washed and soaked rice. Let it cook a little with the other ingredients.

5. Add water, or any kind of stock, till it's about one knuckle on your index finger above the level of the rice.

6. Bring to a boil and follow the same instructions as the ones for plain steamed rice (Step 2 onwards). Once the water has reached the level of the rice, you can optionally add finishing spices such as saffron in milk, roasted dry fruits, garam masala, etc., before closing the lid and letting the rice absorb the rest of the water.

7. At the time of fluffing, you can garnish with herbs.

Rice Algorithm

VARIETY RICE	PULAO
Low-amylose (sticky) rice is cooked separately and a flavour base is added to it	High-amylose (like Basmati) rice is cooked along with the flavour base in one shot

VARIETY RICE

Pressure cook/cook rice

↓

Flavour Base

↓

Spice Powders

↓

Cooked rice ←

Salt

↓

Tadka

↓

Garnish

PULAO

Flavour Base

↓

Add soaked/washed rice

↓

Stock

↓

Salt

↓

Let the rice cook

↓

Finishing Spices

↓

Garnish

The Indian Bread Algorithm

In this part of the world, if you aren't eating rice in some shape or
form, you are probably eating wheat in some shape or form. Wheat, the
second most consumed grain in this part of the world, is largely had in
the form of unleavened breads like rotis and parathas, and occasionally
leavened breads like naans, kulchas and khameeri rotis, etc. There is
some fascinating history, involving famines and the British Raj, about
why wheat displaced millets as a staple in many parts of India over the
last century, but we will skip that for now. As we learnt before, whole
wheat flour in this part of the world is made slightly differently. Here,
atta tends to be ground in a traditional mill called a chakki, which does
a fair bit of mechanical damage to the grains, resulting in some amount
of cooked starch and damaged protein. By damaged, I mean that it's
not going to form the kind of stable gluten structures required to bake
leavened bread. What atta is perfectly designed for is to make rotis and
parathas that are soft and flaky, as opposed to being chewy. Anyone
who has tried to make chapattis with maida will attest to this fact. Maida
has better gluten-forming capabilities, which is why it's preferred for
making leavened breads like naan, kulcha, or baking cakes and breads.
It doesn't work for chapattis because when you add water to maida,
it forms a stretchy, strong gluten structure that we do not want in a
chapatti or paratha.

For those who live in the West, there are two other kinds of wheat flours
commonly available: All-purpose flour and bread flour. There is also
whole wheat flour, which might tempt you to think that it is like atta,
but it is not. Yes, it's flour that includes some of the bran and germ, but
since it has not been chakki-ground, it is not suitable for chapattis.
It will result in a chewier and more fibrous product. All-purpose flour
is similar to maida, except it has a higher protein content that makes it
more suitable for baking purposes. Maida is more finely milled, has less

protein content and tends to be suitable for both baking cakes as well as naans and kulchas, which don't need the structural strength that a cake or bread does. If you want to turn maida into all-purpose flour, simply add 5 g of vital wheat gluten to every 100 g of maida. This will make it more suitable for baking bread.

Gluten Breads

So, let's get going with the algorithm for an unleavened wheat-based bread dough.

1. Take atta and water in a 1:1 by weight ratio and roughly mix them together. Let this sit for 30 minutes. Thanks to a process called autolysis, the wheat will mix with water and form gluten structures by itself, with no kneading required. This is a 100 per cent hydration dough. Depending on the brand of the atta, the room temperature and the alignment of Beta Centauri, this can either become a sticky dough that is hard to handle, or a slightly hard dough that will result in papad, not chapatti. If you are a beginner, you might want to start with less water and work your way up once you get used to managing a moderately wet dough. More water results in a softer final product.

2. Now, gently work salt into the dough till it's evenly distributed. As a general rule, you can use 1% salt by weight of the dough as a rough guide. Use more if it's not seasoned well enough for your taste. Salt is not added during the autolysis phase since it tends to tighten gluten, and we do not want that till the dough forms an extensible structure that gives us a perfectly soft, yet structurally sound, chapatti. If you want an even softer chapatti, you can use boiling water. Personally,

I prefer my chapattis to offer a modicum of resistance to my teeth and not melt in the mouth. So, experiment with both these variables—the temperature and amount of water—till you get the kind of dough you like.

That's it. No kneading for long periods of time required. That said, if you still want to knead the dough, go ahead. It won't hurt, other than your deltoids and bicep muscles.

If you are making a leavened bread, you have two choices. Using a chemical leavener, like baking powder/baking soda, or a biological leavener, like yeast.

The general way to do this with yeast is:

1. Take maida and water in a 1:0.7 ratio. You can substitute up to 50 per cent of the maida with atta if you want a more whole-wheaty flavour. If you do this, remember to add a little more water because atta is a thirstier flour. Let it sit for 30 minutes to autolyse.

2. Work salt into the dough and then add half a teaspoon of instant yeast. The more you add, the faster the fermentation will be. Cover the dough with a wet cloth and let it rise till it doubles in size. Depending on where you live and what time of the year it is, this might take anywhere from 30 minutes (in Chennai) to 2–3 hours (in Srinagar). Place it inside a switched-off oven to accelerate the process.

3. If you want to take your naans and kulchas to the next level, place the kneaded dough in the refrigerator to let it rise more slowly over a longer period of time. Longer fermentation always results in richer and more complex flavoured bread. Do not forget to cover the vessel or else the dough will dry

out in the fridge. If you are fermenting overnight, or for twenty-four hours, all you need is flour, water, salt and yeast. If you are using a more rapid room temperature rise, adding some enriching ingredients like milk, eggs, fat or sugar will improve the flavour and texture of the final product.

If you are using a chemical leavener, there are two options: baking soda and baking powder. Baking soda is sodium bicarbonate and does nothing till it meets an acid. Baking powder is sodium bicarbonate mixed with a dry acid. It is ready to rock and roll the moment water is introduced to the party. If you are not someone who bakes cakes regularly, don't buy baking powder because it has a shorter shelf life than baking soda. Follow the same instructions and replace yeast with baking soda. You will need to add yoghurt to the dough as well in step 2. Adjust the amount of water suitably since yoghurt is also mostly water. Finally, don't let it rest for more than 30–45 minutes because chemical leaveners can be fast and do not depend as much on room temperature.

Once you have these two doughs, there is no science per se to rolling it out into a perfectly round chapatti, naan or kulcha. All it takes is practice. It's the kind of craft that requires either personal mentoring by an expert or just years of individual effort. In my experience, a non-circular chapatti tastes just as good as a perfectly shaped one, so don't waste too much time on geometry.

General Dough Tips

1. If you are planning to deep-fry (like a puri or kachori), the dough needs to have much less water, so add more fat instead. Fat shortens gluten strands, which is what you need in a puri that needs to be flaky without being chewy.

2. Use as little of dusting flour as possible (for puris, don't use any). Use oil instead. The dusting flour will burn in the 170°C oil and turn into not-very-nice-to-eat compounds.

Non-Gluten Breads

Breads can be made from rice, corn, lentil and millet flours, but because these flours do not contain gluten-forming proteins, they are hard to roll out without cracking and tearing. There are two traditional ways of solving this problem. One of them involves being a grandmother who has done it all her life, so it comes down to experience and muscle memory of dealing with non-gluten-based doughs. But what you can do to make life easier is to use really hot water to gelatinize the starches. Remember our enemy amylopectin, the sticky starch? We don't like it much in our rice, but it is our friend when it comes to flours. Using really hot water to knead the dough will make it sticky and relatively easier to roll out. Bear in mind, it still won't be as easy as chapatti dough.

The second method involves the use of third-party starch binders. You can use 30 per cent atta or maida, along with 70 per cent non-gluten flours, to take advantage of wheat's unique stretching properties. If you don't want to use wheat at all and want the full flavour and nutrition of the non-gluten flour you are using, the modernist solution to the problem is xanthan gum. It's a surreally powerful thickener that adds no flavour of its own. The best part is that you need a really tiny amount. In fact, a small pinch of xanthan gum can help you make millet rotis that do not break up like the Balkans.

Indian Bread Algorithm

Glossary of Modules

Hydration	Slow addition of water or watery liquids like milk or yoghurt while kneading to reach a desired level of softness
Fat	Adding fats to flours impairs gluten formation and lends a flaky texture
Salt	In all forms – table salt, black salt, soy sauce etc
Leavening agent	Yeast or Baking powder or Baking soda (Baking soda requires an additional acid like yoghurt)
Binding agent	Xanthan/Guar gum, Atta/maida (for gluten-free breads)
Spices + Vegetables	Powdered/fresh spices & raw/cooked vegetables mixed into the dough or used as stuffing

Rest	Letting the dough sit for ~30 mins (for unleavened breads) and longer if leavening agents are used
Roll	Dividing and rolling dough into the desired shape of the bread
Create layers	Creating layers in dough separated by fat
Tawa cook	Cook one side, flip, smear with fat, flip, smear with fat, flip
Flame Grill	Grill directly over a flame
Deep fry	Fry in oil at 175°C
Baste	Smear fat after the bread is cooked

Atta-based unleavened flatbreads						
Chapathi	Salt → Water → Rest → Roll → **Tawa cook** → Flame grill → Baste					
Plain Paratha	Salt → Yoghurt + Water → Rest → Roll → Create layers → Roll → **Tawa Cook** → Baste					

Atta-based unleavened flatbreads						
Stuffed Paratha	Salt → Yoghurt + Water → Rest → Roll → Stuffing → Roll → **Tawa Cook** → Baste					
Thepla	Salt → Fenugreek + Coriander leaves → Turmeric, red chillies, ginger garlic paste, coriander powder → Fat → Yoghurt + Water → Rest → Roll → **Tawa Cook**					

Fried unleavened flatbreads						
Poori(atta)/ Luchi(maida)	Salt → Semolina → Fat → Water → Rest → Roll → Deep fry					
Kachori	Salt → Fat → Water → Rest → Roll → Stuffing → Roll → Deep fry					

Leavened Flatbreads (Maida or Atta)						
Naan/ Kulcha	Salt → Yoghurt + Water → Leavening agent → Water → Rest → Roll → **Tawa Cook** → Flame grill → Baste					
Bhatura	Salt → Semolina → Fat → Leavening agent → Water → Rest → Roll → Deep fry					

Leavened Flatbreads (Maida or Atta)						
Jowar Bhakri, Bajra Roti, Akki Roti, Oratti	Salt → Binding agent → Herbs+ Spices → Warm water → Rest → Roll → **Tawa Cook** → Baste					

	Bread	Hydration	Autolyse (mins)	Enrichment (optional)	Leavening
Atta	Chapatti	100 per cent water	30	None	None
	Paratha	100 per cent water	30	Yoghurt, oil	None
	Puri	100 per cent water	0	Oil	None
	Khamiri roti	100 per cent water	30	Sugar	None
Maida	Naan	70 per cent water + yoghurt	30	Milk	Yeast/ Baking soda
	Kulcha	60 per cent water + yoghurt	30	None	Yeast/ Baking soda
	Bhatura	60 per cent water + yoghurt	30	None	Yeast/ Baking soda

The Indian Gravy Algorithm

So, now that we've covered staples—rice and wheat—it's time to start metamodeling the great Indian gravy side dish, the one whose spectacular diversity the West tends to reduce to a word referring to an orangey-red sauce of British origin with things floating in it—curry. Remember the butter chicken we spoke about earlier in this chapter? The one where a professional kitchen is able to put it together in under 10 minutes? So why does it take us several hours at home? The reason is that restaurants use industrial methods that originated 300 years ago in France. French cooking was among the first cuisines to adapt to the Industrial Age. It was not only the first to document standardized methods and recipes for dishes, but it also went a step ahead and documented the preparation of building blocks for dishes that could be made in bulk, and then actual dishes could be assembled in very little time.

French cooking tends to use sauces, or base gravies, that can be made ahead of time. So, the actual cook time of a dish is very little. And this is not merely a productivity hack. By starting with a base sauce, the chef can layer additional flavours on top of the base. If you recall the flavour layering principles from Chapter 2, great dishes are constructed by layering flavours such as base flavours, acids, finishing spices and so on. The French school has been influential in being the blueprint for how any fine dining restaurant's kitchen runs.

This approach is quite common in Indian restaurants as well, but not in the Indian home. The reason is simple—home-cooking methods evolved at a time when refrigeration was largely absent. For instance, my home got its first refrigerator in the mid-1980s and, even then, my mother stored just milk, yoghurt, idli batter and little else in it. But devices don't change behaviour overnight. It took decades for urban families to consider it okay to store leftovers. Most food in India is cooked and eaten fresh, which is why there tends to be extraordinary focus on making just the precise amount, so that wastage is minimal. In the pre-refrigeration era, this was crucial because food spoils much faster in tropical conditions. We all know that fungi and bacteria absolutely love temperatures above 30ºC, a temperature considered to be early winter in Chennai, where I live.

With the use of refrigerators increasing over time, you will find giant 700-litre fridges now commonly available in urban stores. In fact, there is a rather interesting aspect to the design choices fridge manufacturers make for India. Freezer space tends to be much smaller because we don't store or consume meat in large quantities, frozen vegetables are still a tiny market, and more crucially, a large number of people still haven't realized that freezing is the absolute best way to store literally anything long-term. For a country obsessed with not wasting things, that we don't put the freezer to better use is an absolute travesty.

For now, let's get our gravy algorithm going by starting with an example. The simplest dal you can make likely involves the following steps:

1. Pressure-cook lentils.

2. Add salt.

3. Temper with whole spices like mustard, chillies and cumin.

4. Garnish with fresh herbs like coriander leaves.

This is perfectly fine if you are making a more elaborate main dish, so you want the side dish to be non-overwhelming. However, if you are planning to have this with plain steamed rice, you might want to consider a more flavour-heavy dal, for which the method might look more like this:

1. Heat fat and add whole spices.

2. Cook onions till they brown (remember the Maillard reaction from Chapter 3).

3. Add ginger, garlic and tomatoes. Let them reduce and thicken.

4. Reduce heat and add spice powders, such as coriander and cumin powder. Remember, spice powders are sensitive to high heat.

5. Add pressure-cooked lentils.

6. Add more water to achieve the desired level of consistency.

7. Add salt and finishing spices, such as garam masala.

8. Turn off the heat, temper with whole spices like mustard and cumin.

9. Garnish with herbs and squeeze some lime juice (recall the balancing with acid concept from Chapter 4).

This will yield a decent dal tadka. If you take this same set of instructions, replace lime juice with tamarind water as the acid, and use a slightly different mix of spice powders, you will get a passable sambar.

1. Heat fat and add whole spices.

2. Cook onions till they turn brown.

3. Add vegetables and tomatoes. Let them reduce and thicken.

4. Reduce heat and add sambar spice powder mix.

5. Add tamarind water (as the acid).

6. Add water to achieve the desired level of consistency.

7. Add pressure-cooked lentils (toor dal).

8. Add finishing spices (some more sambar powder or finishing sambar spice mix).

9. Turn off heat and temper with whole spices like mustard and cumin.

10. Garnish with herbs, such as coriander leaves.

All we did here was add the acid earlier, because as we learnt in Chapter 4, tamarind needs a longer time to cook while lime juice must not be exposed to sustained heat and is best added at the end.

We can generalize this a bit and arrive at this:

1. Heat fat and add whole spices.

2. Cook onions till they brown.

3. Add ginger and garlic and let it cook.

4. Add tomatoes, and let it reduce and thicken.

5. Add prepped ingredients (vegetables/lentils/meat/seafood).

6. Reduce heat and add spice powders.

7. Add more water or stock (vegetable/meat/seafood) to achieve the desired level of consistency.

8. Add finishing spices.

9. Turn off heat and temper with whole spices.

10. Garnish with herbs.

You can, of course, skip or swap any ingredient mentioned here using the tips and tables from the previous chapters. For instance, skip the onion and garlic if you believe some random alignment of celestial bodies will find your allium consumption displeasing.

But, as you might have noticed, this algorithm will only get you a red-coloured tomato/onion/garlic-based gravy, which isn't the only game in town. So, before we generalize our algorithm a bit more, let's head back to our snooty friends in the French kitchen. Remember how they use prepped base sauces? In the nineteenth century, Marie-Antoine Carême (the last name is pronounced best while attempting to dislodge some phlegm in your throat) anointed these five sauces as the building blocks for all of French cuisine in his monumental work *L'Art de la Cuisine Française au Dix-Neuvième Siecle*:

1. Béchamel: A white sauce made from butter, flour and dairy.

2. Velouté: A white sauce made from butter, flour and white stock (poultry, sea food or vegetable).

3. Espagnole: A brown sauce made from butter, flour and brown stock (red meat).

4. Tomato: A red sauce made from butter, flour and tomatoes.

5. Hollandaise (added later): A sauce made using egg yolks, clarified butter and acid.

The critical thing to remember here is that while you can start your cooking process by making these sauces fresh, refrigerating them in bulk allows you to begin with one of these sauces and then adding your own flavours to it. The idea is to not use this base sauce as the only flavouring. Adding your own flavours creates a layered experience where the well-integrated, muted flavours of the sauce form the base, while the brighter, fresher flavours sit on top. The overall experience is better than making the whole thing from scratch, in which case there are no base flavours, just a ton of fresh ones.

Restaurants in India tend to make a few standard base sauces every day and then use them when an order comes in. Butter chicken, you said? No problem. Grab some tandoori chicken, ladle out some makhani gravy, add some fresh flavouring in terms of spices, add some acid for contrast and dollops of butter or cream to smoothen everything out, and get it to the table in under 10 minutes.

Base Gravies

It's now time to introduce you to the idea of making base gravies for Indian cooking. Let's start with what makes a gravy evoke a specific flavour from a region, cultural label or cooking style. These are just broad-brush generalizations, by the way.

1. Gujarati: Use of gram flour, yoghurt, sesame seeds, sugar and lime juice.

2. Punjabi: Use of ginger, garlic, onion and tomato-based gravies with coriander and cumin powder.

3. Chettinadu: Use of shallots, garlic, curry leaves, red chillies and fennel (saunf).

4. Malabar: Use of shallots, garlic, curry leaves and coconut milk.

5. Mughlai: Use of cashew nut paste, or cream, and garam masala (mace, nutmeg, cardamom).

6. Bhuna: Use of browned onions and slow cooking to thicken a gravy till it coats the main ingredient.

7. Banarsi: Use of spices mixed in yoghurt.

8. Jain: Use of turmeric, asafoetida, coriander, jeera and chilli powder (no onion or garlic).

9. Bengali: Use of mustard oil and panch phoran spice mix.

What emerges is a broad set of patterns for making side dishes unique to a culture or region. We are not aiming for authenticity here. Authenticity in food is but a silly idea. Food has continuously evolved in every single household, with every single meal. What we are trying to do is to arm ourselves with a metamodel on how to concoct a recipe that evokes a specific region and culinary tradition, not necessarily to do a better job than a person who has grown up in that culture.

There are three building blocks for our modernist gravy dishes:

1. Base gravies.

2. Spice combinations (both for use at the start and end of a dish).

3. Choice of fat and flavouring oils.

Let's start with the ubiquitous makhani gravy that can be used to make butter chicken, paneer butter masala, veg makhani and dal makhani. What we want in a base gravy is body and mild, well-integrated flavours, which means cooking for a long time. It's best to make a big batch over the weekend.

Here's how you can make makhani gravy:

1. Take a big pot and heat butter and oil in it.

2. Add coarsely chopped onions, ginger, garlic and tomatoes. Use more tomato than the onion.

3. Add whole spices like Kashmiri red chillies, black cardamom, cloves, bay leaves and pepper.

4. Add thickening agents like poppy seeds or cashew nuts.

5. Add water (or any kind of stock).

6. Optionally, add cream (it's better to add it fresh when you are making the dish).

7. Let it cook at medium–low heat for at least an hour.

8. Once it cools down, blend it in a mixer. Strain out all the fibrous husks of the spices, pour the gravy in silicone cups and freeze them. When you are making a dish, take as many cups as you need, microwave them to bring them to cooking temperature and add to your dish.

We can also make a Chettinadu-style base gravy:

1. Heat sesame oil.

2. Add ginger, garlic, shallots, curry leaves, red chillies and fennel.

3. Add chicken stock and let it cook for an hour.

4. Add rice flour as a thickening agent.

5. Blend it, strain it and freeze.

Or a Malabar-style base gravy:

1. Heat coconut oil.

2. Add garlic, shallots, curry leaves, red chillies and coriander seeds.

3. Add coconut milk diluted with water or stock.

4. Let it cook on low heat for 30 minutes.

5. Blend it, strain it and freeze.

Or a Mughlai-style base gravy:

1. Heat ghee and add a puree of onions, ginger and garlic to it.

2. Add whole garam masala spices: mace, clove, cardamom and nutmeg.

3. Add thickening agents like cashew nuts and poppy seeds.

4. Add milk and let it cook on low heat.

5. Blend it, strain it and freeze.

I'm sure you get the drift. The idea is to use a regional or culture-specific choice of fat and spice combinations to create a base sauce with thick consistency, which you can add to your eventual dish.

Gravy (heat)	Fat (add)	Ingredients (once cooked, add)	Stock (Cook, then)	Enrichment
Generic north-indian	Ghee	Onions, ginger, garlic, tomatoes, coriander powder, cumin powder, turmeric, chilli powder	Water	
Makhani	Butter	Ginger, garlic, coriander powder, cumin powder, lots of tomatoes, turmeric, chilli powder	Water	Cashew paste, cream
Mughlai	Ghee	Pureed ginger, garlic, onions, mace, nutmeg, cardamom, poppy seeds	Milk	Cashew paste
Chettinadu	Sesame oil	Shallots, curry leaves, fennel, garlic, ginger, whole red chillies	Chicken stock	Rice flour
Malabar	Coconut oil	Shallots, curry leaves, garlic, coriander, whole red chillies	Diluted coconut milk	
Bengali	Mustard oil	Mustard, nigella, ajwain, cumin, fenugreek onions	Seafood stock	

Base Gravy Tips

1. In general, avoid adding salt or sugar to your base gravies. You can add them when you make the dish and have better control over the final flavour profile.

2. Always strain the gravy after blending to get rid of all the fibrous husks. They won't have any flavour left after an hour of cooking.

3. We aren't looking to maximize the Maillard browning reaction in a base gravy because we will do that when we make the dish.

Spice Mixes

If you remember our chapter on spices, you will recall that buying powdered spices is not a great idea. This is because they lose flavour at a very rapid rate once opened. It's better to make small batches of region or culture-specific spice mixes and use them instead. Of course, it's not like people in a particular region use the exact same spice mix every single time. After all, no one wants the same flavours every single day. These combinations, however, will help you pick combinations that we know work because they have stood the test of time.

Spice	Flavour Molecule	Flavours	Pairs with
Cinnamon	Cinnamaldehyde Linalool Eugenol Caryophyllene	Warm, spicy, pungent Floral, woody, spicy Medicinal, woody, warming Woody, spicy, bitter	Cumin Cardamom Clove Pepper
Clove	Eugenol Caryophyllene	Medicinal, woody, warming Woody, spicy, bitter	Cinnamon, Fenugreek Pepper
Star Anise	Anethole Cineole Phellandrene	Sweet, liquorice, warming Medicinal, penetrating Green, peppery, citrus	Nutmeg, mace, cinnamon Black cardamom, ginger Pepper
Fennel	Anethole Limonene Pinene	Sweet, liquorice, warming Medicinal, penetrating Green, peppery, citrus	Nutmeg, mace, cinnamon Cardamom Pepper, cumin
Nutmeg	Myristicin Geraniol Eugenol Sabinene	Woody, warm Rosy, sweet Medicinal, woody, warming Citrusy, peppery, woody	Ginger, pepper Curry Leaf Cardamom, clove Garlic

Spice	Flavour Molecule	Flavours	Pairs with
Mace	Sabinene	Citrusy, peppery, woody	Garlic, pepper, curry leaf
	Terpineol	Floral, citrusy	Coriander
	Safrole	Sweet, warming	Star anise
	Eugenol	Medicinal, woody, warming	Clove
Shahjeera	S-Carvone	Spicy, menthol	Star anise, cinnamon
	Limonene	Citrus, herby	Cardamom, pepper, ginger
	Sabinene	Citrusy, peppery, woody	Nutmeg, mace
Coriander	Linalool	Floral, citrusy, sweet	Cardamom, nutmeg, mace
	Limonene	Citrus, herby	Ginger, shajeera
	Pinene	Woody, spicy	Pepper
	Cymene	Fresh, citrus, woody	Cumin
Cumin	Cuminaldehyde	Earthy, herby	Cinnamon, cardamom
	Pinene	Woody, spicy	Pepper, black cardamom
	Cymene	Fresh, citrus, woody	Coriander, ajwain, nigella, star anise

Tempering Templates and Infused Oils

Tempering is a unique Indian technique to add texture and finishing flavour to a dish by heating spices at a very high temperature, thus muting their flavours and ensuring they don't overwhelm the dish. The general idea is to add both a mild flavour and crunchy texture to a dish. A typical tempering template for Tamil Nadu is mustard, jeera, urad dal, curry leaves and asafoetida in heated sesame oil. Likewise, fennel, carom seeds, nigella, fenugreek and cumin in mustard oil makes for a quintessential Bengali tempering. You can elevate any dish by picking a set of whole spices and fat appropriate to its region and culture. You can also pick some legumes for crunch and red chillies for heat to make your own tempering mix.

Infusing oils is a technique borrowed from East and South East Asia. The idea is to let the spice flavours steep into warm oil over several hours and then use small amounts of that oil to finish a mildly flavoured dish (like a plain dal) with a burst of intense flavour.

Here are some infused oil recipes to get you going:

1. Garam masala oil: Pour hot ghee over whole garam masala spices such as black cardamom, mace, nutmeg, cinnamon, clove and cardamom. Let it infuse over several hours. Filter the spices out.

2. Chettinadu oil: Pour sesame oil over shallots, curry leaves, fennel and garlic.

3. Bengali oil: Pour mustard oil over the panch phoran mix.

4. Sambar finishing oil: Pour ghee over red chillies, fenugreek, coriander and pepper.

The Gravy Algorithm

So, now that we have our building blocks, here is the modernist Indian gravy algorithm, combining all the things we have learned so far. Begin by asking yourself these questions:

1. What kind of gravy do you want to cook? Vegetables/meat/legumes/egg/seafood?

2. What style do you want to cook it in? Punjabi/Bengali/Chettinaadu/Maharashtra, etc.?

3. Prep ingredients (check the section on prepping of ingredients in this chapter): Main ingredients, base spice mixes, finishing spice mixes, base whole spices, tempering whole spices, garnish, fat of choice, acid(s) of choice, base gravy of choice, flavoured oils of choice.

4. Heat your choice of fat and add base whole spices.

5. If your dish involves onions, cook them as appropriate (translucent, light brown, dark brown).

6. If your dish involves ginger or garlic, lower the heat and cook them.

7. Optionally, add a splash of alcohol to deglaze the pan and extract more flavour from the spices.

8. If your dish involves tomatoes, add them. You can also add some ketchup, or tomato paste, and let it reduce.

9. Add the base gravy of your choice and let it cook briefly (it's already cooked, so don't cook it for too long).

10. Add the main prepped ingredients.

11. Lower the heat and add the base spice mix.

12. Add the acids (tamarind or vinegar).

13. Add stock (water/stock/coconut milk/yoghurt) with starch.

14. Add sugar and salt, and taste to check for balance.

15. Bring it to a boil.

16. Switch off heat. Optionally, add finishing spice mix.

17. Adjust for thickness and flavour intensity by adding butter/cream/thickeners.

18. Optionally, add a finishing acid like citrus juice.

19. You can also add umami ingredients like MSG, mushroom powder.

20. Temper using regional, dish-appropriate spices and choice of fat.

21. Add a garnish of your choice.

22. Optionally, add a flavouring oil of your choice.

If this seems overwhelming, it's meant to be. Let's process it in a visual way now.

Finishing Flavours (optional, pick 2-3 things, not all!)	· Tempering and garnish - style specific · Finishing acids - Lime juice · Intensity adjustment - Cream/butter/ghee · Finishing spices - Roasted and powdered spices · Umami - MSG, mushroom powder · Finishing oils - style-specific · Smoking - Ghee on hot charcoal or liquid smoke
Seasoning	· Salt (table, kala namak, soy sauce), sugar (jaggery, brown sugar, white sugar)
Base Acid	· Amchoor, tamarind, vinegar (optional if tomatoes are involved)
Stock (pick one)	· Water (most common) · Vegetable, meat or seafood stock for a richer flavour · Diluted coconut milk
Prepped main ingredients	· If legumes, pressure cook · If vegetables, par-cook and brown in fat · If meat, brine, marinate and brown in fat
Base gravy	· Style specific base gravies to add the background flavour
Alcohol (optional)	· Deglazes pan and extracts more flavours from spices
Flavour Base	· Fat + whole spices · Onions, ginger, garlic, tomatoes (optional) · Chillies (optional)

As you can see, you don't have to follow every single step for each dish! For each block, pick the steps that are appropriate. Some blocks are entirely optional. It's all about layering flavours to the point where you like the taste of your dish, while keeping in mind the role the dish plays in the entire meal. If every dish is intensely flavoured, it does not make for a balanced meal. Here is a practical example to make a fantastic dal makhani:

Finishing Flavours (optional, pick 2-3 things, not all!)	· Garnish - Coriander · Intensity adjustment - Cream and butter · Finishing spices - Garam masala · Finishing oils - Garam masala- flavoured butter · Smoking - Ghee on hot charcoal, liquid smoke or ghee on hot cinnamon stick
Seasoning	
Base Acid	· Salt, jaggery
	· Not required since we added tomato paste
Stock (pick one)	· Dal stock from the pressure cooker · Additional water, if required
Prepped main ingredients	· Black urad and rajma soaked for 8 hours and pressure cooked with a pinch of baking soda
Flavour Base	· A splash of brandy or rum to deglaze the pan
Flavour Base	· Oil + butter · Grated onions till Stage 4 browning · Ginger, garlic paste · Tomato puree + tomato paste · Kashmiri chilli powder and turmeric

The gravy cheat sheet:

1. A good gravy is about striking the right balance of flavours in each layer. If you turn the dial to 11 and use a base gravy, fresh spices, finishing spices and a flavouring oil, your dish will be unpalatably over-spiced. I'd say use finishing spices very sparingly and pay close attention to whether you want to use garlic as a paste or roughly chopped. Consider using finishing oils only in lightly spiced dishes.

2. In general, use whole spices early in the cooking process and spice powders later. However, it is not uncommon to add turmeric and chilli powder early in the process because they are primarily used for colour and heat, not flavour.

3. Remember the rules of flavour and always create a three-way balance between salt, sweet and sour to create a memorable dish. These three will elevate the flavours of the spices you add.

4. If you need to thicken your gravy, add starch-based flours like rice, corn or wheat flours, but remember that these will mute the intensity of flavours in your dish. Consider using xanthan or guar gum, as these modernist thickeners work tremendously well in small quantities and do not add any flavour of their own.

Universal Gravy Algorithm

Basic North Indian	Tadka + onion, ginger/garlic paste, green chillies, tomato
Makhani	Butter + ginger/garlic paste, tomato, onion, tomato sauce, soaked cashews (grind everything to a paste post cooking)
Bhuna	Tadka + fully browned onion, ginger/garlic paste, green chillies, tomato (grind everything to a paste post cooking)
Kurma	Tadka + shallots, ginger/garlic paste, fennel, coconut milk (grind everything to a paste post cooking)
Malabar	Tadka + shallots, ginger/garlic paste, tamarind, tomato, coconut milk
Chettinaadu	Tadka + shallots, ginger/garlic paste, fennel, coconut milk (grind everything to a paste post cooking)
White	Yoghurt, blanched cashew and melon seeds, boiled onions
Salan	Sesame seeds, cashew nuts, peanuts, desiccated coconut, black pepper
OTM	Chopped onions, tomatoes, ginger, green chillies
Rezala	Whole garam masala, ginger garlic paste, boiled onion paste, coriander powder, yoghurt, almond paste and saffron

5. If your gravy is too intense, you can add fat (ghee, butter, cream or coconut milk) to balance it.

6. If your dish feels too heavy and fatty, you can add an acid to reduce the perception of greasiness.

The Chutney and Raita Generator

One of the subcontinent's greatest contributions to the culinary experiences of the world is the chutney. It has different names in every part of India, but 'chutney' happens to be the most commonly used term. Indian languages often use different terms depending on the presence or absence of specific ingredients. Our aim is to build the most general metamodel for taking a bunch of ingredients and blending them into a fine or coarse paste, and optionally tempering it with whole spices and dropping in some acid to balance all flavours. In that light, a raita and chutney belong to the same neighbourhood, just like thuvayal or pachadi.

So, after researching this category of side dish from almost every part of India, a pattern emerges. A chutney involves eight elements:

1. Cooked ingredients: Essentially, these are things that don't taste good raw. It could be beetroot, eggplant, or cabbage for that matter. Pick one of your choice.

2. Raw ingredients: Things that taste good raw, such as carrot or radish, and most fruits. Again, pick one.

3. Nutty ingredients: Like roasted lentils, cashews, coconut, etc. These add body and crunch to the chutney. Pick one or two from these.

4. Herbs: The green things, like mint and coriander. Pick one or both.

5. Seasoning: Basically salt and sugar for balance. Black salt can also add a lovely flavour dimension. You can also use honey instead of sugar. If you are using fruits, you don't need any additional sugar because the fruits will bring fructose to the party.

6. Heat: Use pepper, ginger, or red and green chillies. Pick just one or two from this list.

7. Acid: You can use anything from vinegar to citrus juice to amchoor to yoghurt (which makes it a raita) and tamarind juice. Pick one or two from among these.

8. Tempering: This is optional, and as per regional or dish-specific preferences.

Blend together and temper based on your preference

Cooked ingredients (Pick 1-2)	Raw Ingredients (Optional)	Nutty stuff (roast in oil)	Herbs	Salt	Acid	Sweet
Beetroot	Carrot	Cinnamon	Coriander	Salt	Vinegar	Sugar
Eggplant	Radish	Urad	Mint	Kala Namak	Lime juice	Jaggery
Dates	Pomegranate	Peanuts	Chives		Yoghurt	Honey
Tomato	Raw Mango	Sesame	Basil		Sour Cream	
Cabbage	Ginger	Niger seed	Lime leaves		Tamarind	
Tamarind	Plums	Chana dal			Orange juice	
Capsicum	Pineapple	Cashews			Pineapple juice	
Garlic	Amla	Macadamia				
Onion	Raw Papaya	Walnuts				
Chillies	Dried Seafood	Brazil nuts				
Sweet potato						

Let's take an example.

- Cooked ingredient: Beetroot

- Raw ingredient: Carrot (because it goes well with beetroot)

- Nutty ingredient: Grated coconut

- Herbs: None

- Seasoning: Salt. Beetroot has enough sugar.

- Heat: Red chillies, because we want to keep the theme in the red to purple department.

- Acid: Yoghurt

- Tempering: Mustard, asafoetida, curry leaves and jeera

There you go! These ingredients will give you a fantastic south Indian-style beetroot chutney.

Chutney and Raita Rules

1. Don't pick too many ingredients or your chutney will taste like nothing in particular.

2. Use flavour-pairing rules for ingredients and combine ones that go well together, like basil and mango.

3. Roast nuts or lentils before using them.

4. If you are making a green chutney that predominantly features leaves (coriander, mint, etc.) don't add the acid till just before serving. Strong acids quickly decolourize leaves and turn them into a dull olive green that looks unappetizing. Yoghurt,

however, is not a very strong acid, so it's okay to mix greenish raitas ahead of time.

5. Season raw vegetables with salt ahead of time. Let it extract the extra water out of them to make them crisp and taste better. Do not do this for leaves, as the salt will make them wilt and lose crunch.

6. If you are planning to use raw onions or radishes, consider pickling them in vinegar ahead of time to tame their sharpness. You could also blanch them if you don't want the strong vinegary taste.

The Salad Generator

One casualty of the maddening urban Indian quest to cook everything to death is the salad. A restaurant will serve a melt-in-the-mouth amaklamatic butter chicken, but the salad will usually be a circular assortment of unseasoned sliced cucumber and tomato, with a two-fifth slice of lemon. Other common Indian salad atrocities involve the cabbage, which has incurred the wrath of the Sinaloa cartel and been shredded and drowned in a tub of mayonnaise (coleslaw), and boiled potatoes, beans and carrot kidnapped and thrown overboard into Lake Mayonnaise by the KGB (Russian Salad), and sliced radish or raw onions with no seasoning. Then there is raw capsicum and large slices of raw carrot that are designed to choke you to death.

Okay, maybe I'm being uncharitable here. There are a few half-decent salads in this part of the world: Kachumber salad and sirkewale pyaaz (vinegar-soaked onions) are not bad, and kosambari/kosumalli is downright delicious, but that's really about it. We've given the world more flavours in food than any other part of the world, but we need to

be humble, swallow our pride and learn to make salads like the West. That said, it can be argued that Indians salads are simple because our gravies are complex, that it's just a way to balance out a meal. But, in my opinion, chopping vegetables and serving them on a plate is, while admittedly simple, taking simplicity too far.

Here's an elegant way to approach a good salad. There are seven elements that go into a balanced salad. Of course, this is just a guideline. Feel free to ignore any elements.

1. Greens

2. Carbs: Optional if you are the type who eats a salad only because it's 'healthy'.

3. Vegetables: These can be cooked, pickled or raw. Anyone who thinks raw radish is a good idea must be shredded and thrown into a tub of mayonnaise. Always season raw vegetables with salt because they taste terrible otherwise. Salted tomatoes and cucumbers taste amazing in a salad.

4. Fruits: These could be fresh or sun-dried.

5. Proteins: These could include legumes, paneer (or any other kind of cheese), tofu, eggs, shredded chicken, cured meats, etc.

6. Crunch: Since a salad does not have an intense flavour profile, it needs to have variation in texture and mouthfeel. Adding nuts, roasted papad or fried onions will make a salad taste and feel interesting.

7. Fancy ingredients: Olives, cheese, etc.

Once you have this mixed, make a dressing. A good salad dressing has six components. The most critical ratio to keep in mind is three parts fat to one part acid. Acid is what makes raw ingredients taste good, but if you remember your high-school botany, water doesn't stick very well to leaves and plant surfaces in general. If it did, plants would drown and die rather quickly, and we wouldn't be wondering how to make a decent salad. Culinary acids are mostly watery (lime juice, vinegar) and thus do not stick to plant matter. To make them stick, we use the same principle that we used in a marinade. Fats have excellent sticking properties, which does two things. They make the dressing stick to your salad ingredients, instead of gathering in a sad, watery pool at the bottom of the bowl, and they coat the tongue and mouth and transport flavours. There's just one minor problem: fats and water don't exactly mix well either. The trick here is to emulsify the fat and watery acid, a process that will create a stable, creamy mix of fat and water, like mayonnaise. Vigorously whisking fats and acid together will create a temporary emulsion that is usually good enough for a salad. If you want the emulsion to last longer, you need a peacekeeping molecule, an emulsifier, to prevent domestic disturbances between the fat and water molecules. Egg yolks, mustard paste and honey are very good emulsifiers. Of course, the food industry uses lecithin, which is the molecule in an egg yolk that acts as an emulsifier. If you are planning to make industrial quantities of salad dressing, get yourself a packet of soy lecithin.

Here's the formula:

1. Fat (three parts): Pick a liquid fat of your choice based on regional preferences.

2. Acid (one part): Vinegar, lime juice, pineapple juice, yoghurt, etc.

3. Salt: Common salt, black salt, soy sauce.

4. Sweet: Honey, sugar, molasses, jaggery.

5. Heat: Chillies, pepper.

6. Spices: Garlic, ginger, spice powders.

You can create, for instance, a Bengali-style salad dressing with mustard oil as the fat, vinegar as the acid, black salt as the salt, sugar, chillies for heat and a pinch of Bengali garma moshla for spice.

MIX TOGETHER AND REFRIGERATE

Greens (Pick 1)	Vegetables (Pick 2)	Fruits (Pick 1)	Protein (Optional)	Crunch (Pick 1)	Fancypants (Optional)
Lettuce	Cucumber	Tomato	Boiled eggs	Walnuts	Cheese
Baby spinach	Capsicum	Apples	Chickpeas	Peanuts	Olives
Rocket	Carrots	Berries	Black-eyed peas	Almonds	
Fenugreek	Onion	Grapes	Sprouts	Raisins	
Parsley	Turnip	Peach	Rajma	Coconut slices	
Coriander	Broccoli	Guava	Paneer	Roasted garlic	
Mint	Cauliflower	Avocado	Tofu	Fried onions	
Basil	Beetroot	Watermelon	Paneer	Roasted papad	
Sorrel	Radish	Mango	Chicken	Corn	
Fennel	Pumpkin	Pomegranate	Cured meats		
Wilted Spinach	Beans	Figs	Dried seafood		

+ DRESS JUST BEFORE SERVING

Greens (Pick 1)	Vegetables (Pick 2)	Fruits (Pick 1)	Protein (Optional)	Crunch (Pick 1)	Fancypants (Optional)`
Lettuce	Cucumber	Tomato	Boiled eggs	Walnuts	Cheese
Baby spinach	Capsicum	Apples	Chickpeas	Peanuts	Olives
Rocket	Carrots	Berries	Black-eyed peas	Almonds	
Fenugreek	Onion	Grapes	Sprouts	Raisins	
Parsley	Turnip	Peach	Rajma	Coconut slices	
Coriander	Broccoli	Guava	Paneer	Roasted garlic	

8 The Biryani

Don't let anyone treat you like upma. You are biryani.

—Anonymous

There is a commonly used trope in science fiction called the Gaia hypothesis. It proposes that life on earth is not merely the sum of ecologically interdependent species of organic matter (plants, animals, insects and microbes) but also includes the massively complex inorganic systems that make the planet liveable in the first place, such as climate systems that regulate global temperature, the salinity of sea water, the levels of oxygen in the atmosphere and the maintenance of liquid water that makes up 71 per cent of our planet's surface. The fact that we have enough oxygen in the air is thanks to photosynthesizing phytoplankton that evolved around 2.5 billion years ago. The hypothesis goes on to say that if drawing the boundary of a living organism around its skin is limiting to the understanding of ecosystems of life, where deer eat grass and get eaten by lions and dying lions return to the soil, drawing the boundary at local ecosystems of life is also limiting. This is because it limits our understanding of how interlocking ecosystems of organic matter affect inorganic systems like sea water and climate. In short, the

Gaia hypothesis proposes that Earth is one big living organism with smaller families serving as cogs in a planet-sized body of organic and inorganic life.

That brings us to one of the subcontinent's greatest culinary inventions, the biryani. It is the apotheosis of craft in the kitchen. It brings together the most aromatic varieties of the subcontinent's staple grain—rice—and life-nourishing protein, but not like two families at an arranged marriage. It brings them together like two companies merging and, to quote several PowerPoint presentations, drives synergies across the board. A good biryani is not just conceptually but also literally layered with multi-dimensional flavours—of the meat that has undergone the Maillard reaction at the bottom of the vessel, the umami of the glutamates in the animal protein, the fantastic aromas of the rich spices coating the meat, the layering of fresh herbs and flavour-transporting fat (ghee), the textural contrast between the perfectly soft yet fiercely independent grains of rice and the crunch of fried onions, not to forget the top layer that blends the incredible complexity of saffron and the floral top notes of kewra water—that make it the Gaia of dishes, a layered living system of rice, meat, spices and fat, a complete meal by itself that requires no side dish. It is the single most consumed dish at restaurants in India. According to Swiggy, in 2019, forty-three orders of biryani were received every single minute of the year.

There has been some pointless debate about whether vegetable biryani is biryani. In my opinion, it is. Nomenclature territorialism is stupid in a country where no two homes use the same recipe for a dish. Insisting that there is no such thing as a vegetable biryani is no different from insisting that sambar powder with cumin cannot make an authentic sambar (yes, this is a thing, in case you are wondering). Personally, I don't like biryani made using boneless poultry either. I think the lack of connective tissue ultimately makes for dry, non-succulent meat, but would it make sense for me to file a lawsuit against anyone making a chicken tikka biryani? We all form emotional attachments to the food we grow up with. Nostalgia, as we learnt earlier, is within gossiping distance of the olfactory cortex that processes the experience of flavour. We love food that evokes memories, Michelin-star quality alone is not enough. A biryani is a layered dish of rice and other ingredients, each of them partially cooked separately and then cooked together at low heat, in an airtight mode, to create the right balance of textural variations and explosion of flavour. A pulao, on the other hand, is a rice dish made by cooking rice and other ingredients in one shot. And even this distinction is purely based on semantic convenience. Kashmiri biryani, for instance, is typically made in one shot like a pulao, without the layering. So, let's stop this business of harassing someone for using a name they like for the food they eat.

In this chapter, we will marshal every single food science trick we have learnt over the last 200 pages or so to make good biryani at home. We will do this meticulously, step by step, from prepping the rice, meat or vegetables to finally layering and dum-cooking to get the final product. Once we build a solid foundational algorithm for this dish, we will explore regional variations and modernist experimental takes that make this dish a constant subject of inter-state supremacism and nomenclature territorialism on social media.

The Rice Layer

As discussed in the previous chapter, there are many ways to make rice: plain steaming, as part of a khichdi, as pulao where it is cooked with other flavouring ingredients and being prepped as biryani. The rice, and not the meat, is the star of the biryani. A biryani with overcooked meat will be tolerated because Indians are used to eating overcooked meat all the time. In fact, we tend to see it as a source of protein and little else. But get the rice wrong and the dish will be an unmitigated disaster. What makes prepping rice for biryani tricky is that it undergoes cooking twice, so it has to be partially cooked the first time, so that it does not overcook when layered and dum-cooked with the meat or vegetables.

Let's start with some ratios. In general, a balanced biriyani uses equal parts of rice and meat. So, if you are using 250 g of meat, use 250 g of rice. If you are using vegetables instead of meat, use the same ratio by weight. The idea here is to partially cook the rice (al dente) so that it gets to its full length and size but is not fully cooked. The rice should have a bite to it.

The Biryani Rice Algorithm

1. Wash the rice thoroughly. We want to evict every surface molecule of sticky amylopectin, which is the enemy of good biryani. Do this four or five times, till the water runs clear.

2. Soak the rice for 20 minutes. Then wash and drain again. Soaking the rice will help it cook more evenly.

3. Take a vessel and add lots of water. Since we will partially cook the rice, after fully submerging it, we don't have worry about the amount of water as long as it is more than three to four times the volume of rice. To this water, it is absolutely critical that you add salt. This is how the rice will get seasoned. If you don't season it well, the biryani will taste flat. Remember, because we are using extra water, not all the salt will get into the rice, so add a little more to compensate for this. A general rule of thumb is to add salt till the water tastes like the sea. If you have never tasted sea water, please travel to a seaside city like Chennai. A good biryani is worth this

effort. You can also add whole spices to the rice if you want to add another layer of flavour, but remember that spices aren't water-soluble and not much is going to stick to the rice. A teaspoon of ghee will help here.

4. If you want a more intense flavour, and this is purely a personal preference, cook the rice in meat, seafood or vegetable stock.

5. Bring the water to a boil. Once the rice reaches its full length but is still raw inside, turn off the heat and drain the rice into an open plate. Let it cool down.

6. When straining the rice, please remove any whole spice husks. While restaurants want to give you visual confirmation of the fact that they are being generous with expensive spices by leaving those flavourless husks behind, you be a nice person and remove them. No one wants to be navigating through a minefield of cardamom husks when focusing on a mouthful of orgasmic biryani. If you are up for it, take a piece of thin cloth and make a small sachet of spices (a bouquet garni, if you will) that you want to use. Drop this into the water. That way, you won't have to painstakingly fish the spice husks out later.

The Protein Layer

The magic trick to keep meat tender and moist, while ensuring it is fully cooked, is not marination, as most people may tell you. The aim of marination is to get the flavours to stick to the surface of the meat, but that alone does not help the meat to stay tender. The key to that is brining. A quick recap: Salt dehydrates vegetables but helps animal tissue retain moisture. This is why we drink water with salt and sugar when we are dehydrated. The salt helps you retain the water you just drank and not lose it to perspiration. Fun fact: Bodybuilders have the exact opposite need. They want their muscles to lose as much water as possible so that they look ripped. Your favourite Bollywood star, working out for a six-pack for his upcoming blockbuster, will most likely be put on a low-salt diet for months on end. In fact, he might even drink distilled water to ensure that it has zero dissolved salts.

Back to our biryani. Brining time will vary depending on your choice of meat. Red meat, as opposed to fish or shrimp, needs to be brined for a longer period of time (refer to the table on Page 184).

Here's how you can go about it:

1. Heat a litre of water (or more, if you are making a large quantity) and add salt (8 per cent of the amount of water) so that it fully dissolves. Let it cool down to room temperature.

2. You can add other flavouring agents too, like ginger-garlic paste, spice powders, etc. Brining will cause salt and the other flavours in the solution to get pulled into the meat.

3. Now immerse the meat pieces into the solution, ensuring that no part is exposed to air. This is crucial to prevent bacterial infections. Remember, meat needs to be frozen for storage, and we are keeping this in the regular section of the refrigerator, which is simply not cold enough to deter meat-loving bacteria.

4. Based on the table on Page 184, calculate the duration for which you need to let it brine.

5. Wash the meat in regular water post brining to get rid of the salty water on the surface.

Now, it's time to marinate the meat. The general rules for a good marinade are:

1. Acids: Use at least two acids, one weak and one strong. Yoghurt and lime juice are the traditional choices.

2. Dry spices: Use spice powders made from whole spices, which have been roasted and freshly ground. It's best to make your own biryani masala using the lessons from Chapter 2.

3. Fresh spices: Use fresh pasted ginger and garlic. The store-bought ones taste terrible thanks to the sodium citrate.

4. Fat: This is crucial. Fat is what helps all the flavour stick to the meat. For the most part, there is enough fat in the yoghurt, but it won't hurt to add a little ghee or oil. If you want a more intensely spiced biryani, you can heat the fat and add it your marinade before adding the yoghurt, so that the spices are cooked and their flavour molecules get dissolved in the acid.

5. If you are using vegetables or paneer, add a little bit of gram flour as a binder, so that the marinade sticks.

In general, long marinations are not recommended. Anywhere from 30 minutes to 2 hours is more than enough.

Once you marinate the meat, it needs to be partially cooked. This is best done in ghee and, depending on the kind of meat, it might take anywhere from 10 minutes (for chicken) to 1 hour (for mutton or beef). It is always important to instruct the butcher to leave some fat on the meat. That is what will make your biryani taste way more delicious than lean meat. The par-cooking process of the meat will also generate a flavourful layer of liquid fat called *yakhni*. Once you are done, strain this fat out and keep it aside for the final step.

Layering and Dum Cooking

We are almost at the finish line now, so let's prepare a little. A good biryani layers not just meat and rice, but also adds textural variations between those layers. The typical additional ingredients used are:

1. Fried onions (birista): It is entirely worth frying your own onions. Store-bought ones go rancid in no time. But frying onions takes time and patience. They will seem to take ages to turn light brown, and then all of a sudden, like Usain Bolt at the 70 m line, summon all reserves of browning agility unbeknownst to novice cooks and turn into elemental carbon. Also remember that the onions will continue to crisp after you take them out of the frying pan.

2. Coriander and mint leaves.

3. Masala milk: This is typically a mix of super-delicate spices and flavouring ingredients, such as saffron, rose water and attar, added to mildly warm milk.

A neat little trick I've seen on an excellent YouTube channel—BongEats— is to layer bay leaves at the bottom of the vessel. This not only keeps the meat from burning, but also adds flavour to the biryani. Layer the meat pieces on top of this and then add the rice. Next you pour in the yakhni. Then goes in a layer of herbs, fried onions and masala milk, followed by another layer of rice and a final layer of herbs, masala milk and fried onions. Now, seal the vessel as well as you can to prevent loss of moisture. Let it cook at low heat for 30 minutes. Turn off the heat after that and let it sit for 15 more minutes. If you recall the tips from the previous chapter, what we are doing is letting the par-cooked rice go through a process called retrogradation, where the starches realign themselves to ensure each grain stands out separately.

There is an alternative way to dum-cook the biryani, in an oven. Cook the biryani at 180°C for about 40–45 minutes, followed by 15 minutes of rest. The longer time duration is to account for the fact that air does not conduct heat as well as metal. I'd argue that oven baking is, in fact, safer because the risk of charring the bottom layer of meat is very low. But it is critical to make sure that you seal the vessel as tightly as possible. This is because moisture loss in an oven is much more rapid, given that heat is applied from all sides.

So, let's recap the biryani project plan, with all the science lessons outlined so far.

The Rice Track

1. Wash and soak rice for 20 minutes. Washing removes surface amylopectin and other chemicals, such as talc, which are used in the polishing process. Also, soaked rice cooks faster and more evenly.

2. Par-cook the rice in excess water and salt, till it has grown to its full size, while being raw at the centre.

3. Drain the rice and let it cool down in an open vessel. Cooling is a function of surface area. The more the surface area, the faster the cooling down will be. Also, the chances of some of the hotter grains continuing to cook all the way through will lessen.

The Biryani Masala Track

1. Dry-roast whole spices such as cardamom, cloves and mace. Heating activates the release of the volatile aroma molecules that make up the flavour of the spice.

2. Grind them to a powder. Chapter 2 has a suggestion for a well-balanced biryani masala, but feel free to invent your own using the principles outlined there. You can, if you are not up to making your own spice mix, get sachets of biryani masala that you can use in one go.

3. You can use the masala when cooking the meat/vegetable layer, and additionally sprinkle it during the layering and dum-cooking process, to add a more intense flavour to your biryani.

The Protein Track

1. Use bone-in cuts of meat with some skin on, particularly in case of poultry. Boneless cuts will become dry and rubbery. Brine the meat in an 8 per cent salt solution for the duration based on the brining table. Brining has two advantages: It gets salt into the meat, which makes it tastier, and, in turn, the salt prevents moisture loss from the muscle tissue during the cooking process, which results in moist and tender meat in the biryani.

2. Wash the brine off and marinate the meat in a combination of yoghurt, biryani masala, ginger–garlic paste and any other spice that you fancy. Add some ghee and lime juice, and let it sit for at least an hour. Go easy on the salt in the marinade because you have already brined your meat.

3. Cook the marinated meat at low heat. Remember that any temperature above 65°C will make the meat tough and rubbery. The idea is to turn some of the collagen in the connective tissues into gelatin, which will ensure the meat stays tender without overcooking the tissues. This process

will result in a delicious, melt-in-the-mouth flavour that is the hallmark of a great biryani.

4. This process will also yield yakhni, a rich, flavourful broth of rendered animal fat. Drain it out and store it because this will give the biryani its unctuous mouthfeel.

The Onion Track

1. Chop onions into small pieces. You will need more onions than you think because deep-fried onions shrink. If you tear up, use a small hand fan to blow away the irritant molecules.

2. Heat oil up to 177oC and drop the onions into it. Don't drop all of them in one go. That will cause the temperature of the oil to drop precipitously and make your onions greasy. Fry the onions in batches till they are just short of dark brown.

The Masala Milk Track

1. Warm milk at the lowest setting in the microwave for 10 seconds. To this, add strands of saffron, kewra water and attar (optional).

The Herbs and Other Accoutrements Track

1. Chop and keep ready other add-ins like coriander and mint leaves, etc.

The Dum Track

1. In a thick-bottomed vessel, use bay leaves as the base.

2. Layer 1: Meat pieces.

3. Layer 2: Half the rice.

4. Layer 3: Half the masala milk, herbs and fried onion.

5. Layer 4: Rest of the rice.

6. Layer 5: Rest of the masala milk, herbs and fried onion.

7. Seal the lid and use one of the many tricks you are familiar with by now to prevent moisture loss (using dough to seal the edges, or aluminium foil between the lid and vessel).

8. At low heat, let it cook for 30 minutes. Switch off the heat after that and let it sit for 15 more minutes. Remember that the rice and meat are already 80 per cent cooked before this stage, so keep the heat as low as you can to prevent any scorching at the bottom. If you smell any burning, turn the heat even lower but wait out the 30+15 minutes. If you are using a heavy bottomed vessel and the lowest heat setting on your stove, it should not burn.

9. Serve to a loved one.

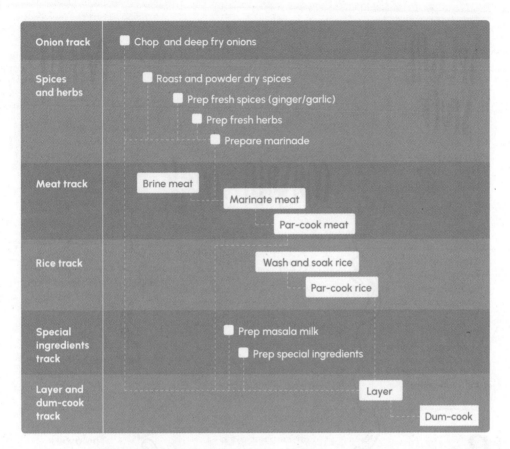

Onion track	☐ Chop and deep fry onions
Spices and herbs	☐ Roast and powder dry spices ☐ Prep fresh spices (ginger/garlic) ☐ Prep fresh herbs ☐ Prepare marinade
Meat track	Brine meat · Marinate meat · Par-cook meat
Rice track	Wash and soak rice · Par-cook rice
Special ingredients track	☐ Prep masala milk ☐ Prep special ingredients
Layer and dum-cook track	Layer · Dum-cook

Regional Variations

Once you've mastered the basic biryani algorithm, you can experiment with a cornucopia of regional variations. The method described in this chapter is closest to the Hyderabadi style. For the most part, the changes will only be in the spice mixes and the way meat is cooked. As long as you follow this multi-track project plan, you can make any kind of biryani. Let's consider a few examples. Again, these are representative recipes, they haven't been personally sourced from the chef of the erstwhile nawab of Awadh or something, so please don't fight with me on social media that my Awadhi biryani is not authentic. Authenticity in food is, and I shall repeat for the millionth time in this book, a silly idea.

SPECIAL STUFF

Awadhi, Hyderabadi
Fried onions

Ambur
None

Thalassery
Fried onions, cashews, raisins, pineapple chunks

Bengali
Fried onions, dried plums, rose water, rose petals attar, fried potatoes, khoya

RICE

Wash and soak for 20 mins for all styles

Thalassery
In ghee, fry onions, spices, add soaked rice, water and salt and let it par-cook

Ambur, Awadhi, Hyderabadi, Bengali
Par-cook in water, salt and bouquet garni of spices

MASALA MILK

Hyderabadi, Thalassery
Fried onions, cashews, raisins, pineapple chunks

Bengali
Saffron, rose water attar

Ambur
None

Awadhi
Saffron, ghee, mace and cardamom powder, rose water

MEAT

Brine meat for all styles

Bengali
Brine meat, marinate in yoghurt and spices, cook in ghee, add water and separate out yakhni and meat pieces

Awadhi
Cook meat in ghee, spices and yoghurt and separate out yakhni and meat pieces.

Ambur
Cook meat in oil, onions, tomatoes, spices and yoghurt, and separate out yakhni and meat pieces

Thalassery
Cook meat in ghee, add yoghurt and coconut milk, and separate out yakhni and meat pieces

Hyderabadi
Brine meat, marinate in yoghurt and spices. Cook in ghee till the gravy coats the meat in a thick paste

Once you've mastered the art of layering biryani, you can experiment a little more.

1. Rice layer: Cook the rice in diluted coconut milk for a fantastically rich flavour. This will work well with a Malabar-style spice mix. You can also experiment with other regional varieties of rice. While some of them are high in amylopectin, the technique used to cook basmati/long-grain rice can be used on most other varieties too.

2. Spice mixes: It's the combination of spices that makes a region's biryani stand out. So, garam masala + saffron + fried onions + mint will give you Hyderabadi-style biryani. Try other combinations as well. For e.g., make a Mexican spice mix using chipotle sauce, cumin powder, garlic and onion powder, and use that for marinating meat.

3. Protein layer: Try different styles of marination using spice combinations from Chapter 2, including experimenting with global cuisines.

4. Herbs layer: Experiment with herbs that combine well and go with your choice of flavours. If you are trying out a Thai-style biriyani, you can use Thai basil and coriander, along with a spice mix that has lemongrass, ginger, shallots, garlic and fish sauce as marinade, and fried shallots for the crunch. When you think of biryani as a canvas, with a template that involves a rice layer, protein layer, region-specific spices, herbs and crunch, the possibilities are endless.

That's all from me, folks! Go forth and experiment in your kitchens, but please don't use your newfound scientific knowledge to harass people who are naturally good cooks. The intent of the science is to help you understand why and how cooking methods work, and how to apply them consistently in other situations. It is not to look down upon someone else's methods. We have been cooking since the dawn of humanity, thrice a day. Understanding food science should be a personal adventure towards more delicious food, not a case of a technical purist schooling Brian Lara on his unconventional shuffle before launching into the most glorious cover drive.

Food science, to paraphrase Albert Einstein, is knowledge that opens up yet another fragment at the frontier of human ingenuity. Three billion years ago, the electromagnetic radiation from a 4.5 billion-year-old thermonuclear reaction (which started in a fireball 150 million km away and, till date, is filtered by Earth's atmosphere to prevent us from being fried to a crisp) was used by a chlorophyll molecule concocted by a cyanobacteria floating in the oceans, which then happened to form a symbiotic relationship with larger organisms to ultimately develop into plants. Billions of years later, a large bipedal ape managed to figure out how to grow these plants to produce grains and vegetables, and then domesticated animals that eat these plants. With the discovery of fire, he invented what is quite possibly the most game-changing thing in history—cooking, where hard-to-digest plant products, filled with evolutionarily designed nasty defence mechanisms, were turned into nourishment that helped this ape develop its brain to the point where it fine-tuned what was just nourishment into the art and science of gastronomy. Today, it lets us sit and ponder with awe at the fundamental interconnectedness of things in the universe while eating morsels of the most perfectly cooked biryani.

Methodology

When you read a scientific paper, often more than half of its contents focus on assumptions, caveats and testing methodology. To the average, impatient reader, it can be rather frustrating if all he/she wants is to quickly get to the summary. And rightfully so. Science works because one person's findings tend to be peer-reviewed by others before they can be cited as a point of reference. And peer reviews require complete transparency on experimental methods and data. You have to declare up front not just what it is that you are testing but also what it is that you are not testing. This is besides describing how you are testing it so that someone trying to replicate your results can use the exact same set-up.

But our kitchens, despite my marketing bluster, are not laboratories in the literal sense. And cooking, which is mostly chemistry, is notoriously subject to local conditions, most of which are very hard to control. Scientists in laboratories will use precision thermometers, distilled water and standardized chemicals to conduct and report reactions they observe. Cooks deal with water in our homes that can vary day to day in terms of dissolved solids and pH levels. The temperature and humidity varies every single day and across latitudes. The tomatoes you use will be sour one day and sweet the next. The rajma (or other beans) you have in your pantry can take 15 minutes to pressure-cook when they are fresh and 30 minutes as they age. The stove you use, depending on the last time it was serviced, will put out different intensities of heat for a given setting. How an induction hob heats your food is very sensitive to the material and thickness of the cooking vessel.

259

So, here is what is important. As much as I might ask you to not trust recipes blindly, you should not trust the temperatures and timings blindly in this book either. They are meant to be a starting point. If those timings and temperatures work for you, that's fantastic, but all it means is that you probably live in south Chennai and buy groceries from the same places that I do.

What is a more useful takeaway from this book is the methods I use to arrive at the conclusions that I do, so that you can apply the same methods in your kitchen.

Journal Everything

If you didn't write down what you did on the day you made the perfect chapatti, you won't be able to replicate it with ease. Let me give you an example.

Date	Brand	Atta (g)	Water (g)	Outcome
18-3-20	Killsbury	400	200	Papad
19-3-20	Usherwhat	400	200	Still Papad
20-3-20	Killsbury	400	300	Kinda okay
21-3-20	Killsbury	400	400	Perfect
21-3-20	Killsbury	400	450	Sticky mess

Just because I said, 'Use 100 per cent hydration for a soft chapatti', don't treat it as the universal verified truth. The behaviour of flours is very sensitive to humidity and temperature, not to mention the milling methods used. Also, 'soft' is a rather subjective feeling. My idea of a soft chapatti also includes some slight chewiness, which is why I tend to prefer slightly more gluten development than others who might prefer a more 'soft and flaky' texture. If that's what you want, you might need to play around with two other variables: the temperature of the water you are using and the addition of a fat (which shortens gluten strands).

So, record what you do in the kitchen and refer to it the next time you make the same dish. You will realize that data is always empowering.

Taste, Texture and Sight, Not Just Time

Every time you hit upon a specific ratio, temperature and time combination that hits the sweet spot for a particular dish, don't just stop at journaling and bookmarking it, and declaring victory. Take the time to build some muscle memory of the texture that you think worked for you. For instance, make it a habit to take a grain of par-cooked rice that you are making for biryani and crush it to see what level of doneness works for you. Also, bite into it so that you know how much rawness at the centre of the grain ultimately makes for a perfect dum-cooked biryani. In fact, a starting point for me was lots of tacit knowledge from people who are natural cooks. In the last few years, anytime I ate something delicious at someone's place, I would talk to the person who cooked it and try to glean the knowledge that rarely gets documented. What makes a pakora particularly crispy? Turns out it's the mixing of some rice flour to the gram flour. Often, the most useful bits of knowledge are the 'do this till it feels like . . .' nuggets from great cooks. When it comes to Indian cooking, there's an interesting socio-historical aspect to a lot of tacit knowledge about texture. Because of the historical taboo against constantly tasting the food one is cooking (because your saliva will come in contact with it) seasoned cooks have evolved a lot of visual and tactile cues to determine cooking milestones. This is often so ingrained in older folks that my mother, for instance, will call me or my brothers to taste what she is cooking for salt, heat (as in, spiciness) and sourness, because she cannot get herself to do it! So, taste what you cook all the time, and keep in mind that hot food tastes milder in intensity. This is why coffee that has cooled down to room temperature tastes more bitter than hot coffee.

A/B Testing

Let's say a recipe instructs you to add a teaspoon of oil to the boiling water in which you are cooking noodles or rice to prevent it from sticking, but you aren't entirely sure if it is necessary. So, try it with the oil once and, on another day, without the oil. See if it made any difference. But don't stop at that because two data points are merely anecdotal. Several other variables might have affected your end result. So, try it a few more times both ways and then average the results. A general principle is to always compare the effect of an ingredient with its absence, not with another ingredient. Likewise, if you are testing a technique, it is better to compare it with the way you would normally do it.

Location, Location, Location

The single biggest variable when it comes to cooking is where you live. The latitude determines the average temperature in your kitchen, which can change a lot of things (from how long it takes for your naan dough to ferment to whether or not coconut oil is solid). It also determines seasons that, in turn, affect the tastes of ingredients, not just at the source but also how you perceive their flavours. The same food tastes more intense at room temperature as opposed to eating it outdoors, where it might be hotter or colder, depending on where you live. Altitude, in addition to affecting temperature, also affects the boiling point of water, which in turn changes a lot of things because water is, well, everywhere.

Tools I Use

1. Instant-read thermometer: This is quite cheap and will truly improve your deep-frying and meat-cooking skills.

2. Microwave timer: For example, I might want to pressure-cook urad dal for 20 minutes, and to do that, I first bring it up to full pressure, after which it blows a whistle and the last thing I want to do is fiddle around with my smartphone because my hands are wet and sticky. Every microwave has a convenient timer that will beep to remind you when the time's up.

3. My wife's distinctly superior sense of aroma and taste: Research tells us that, on an average, women have a keener sense of taste and smell. Getting neutral feedback on the balance of heat, sourness, saltiness and sweetness, in addition to aroma, is critical to improving one's skills.

References

Introduction

Ratio: The Simple Rules behind the Craft of Everyday Cooking, Michael Ruhlman, Amazon, Scribner (2009).

Keys to Good Cooking: A Guide to Making the Best of Foods and Recipes, Harold McGee, Penguin (2013).

'Update on Food Safety of Monosodium L-Glutamate (MSG)', Helen Nonye Henry-Unaeze, 18 September 2017, https://pubmed.ncbi.nlm.nih.gov/28943112/ (last accessed on 25 June 2020).

Chapter 1: Zero-Pressure Cooking

The Food Lab: Better Home Cooking Through Science, J. Kenji López-Alt, W.W. Norton and Company, Inc. (2015), p. 28.

Modernist Cuisine: The Art and Science of Cooking, Nathan Myhrvold, Chris Young and Maxime Bilet, The Cooking Lab (2011), p. 280.

The Feynman Lectures on Physics, vol.1, Addison–Wesley, USA (1963), pp. 2–5.

Essentials of Food Science, Vickie A. Vaclavik and Elizabeth W. Christian, Tad Campbell, Springer Nature (2014), p. 40.

'Materials: Definition of Cooking', *The Food Lab*, J. Kenji López-Alt.

'The Healing Components of Rice Bran', Nurul Husna Shafie and Norhaizan M.E., ResearchGate (2017), https://www.researchgate.net/figure/The-structure-of-a-rice-grain_fig1_324246893 (last accessed on 29 May 2020).

'The Ultimate Pressure-Cooking Chart', FastCooking.ca, https://fastcooking.ca/pressure_cookers/cooking_times_pressure_cooker.php (last accessed on 25 June 2020).

The Science of Cooking: Every Question Answered to Perfect Your Cooking, Stuart Farrimond, Dorling Kindersley Limited (2017), p. 134.

Culinary Nutrition: The Science and Practice of Healthy Cooking, Jacqueline B. Marcus, Academic Press (2013), p. 61.

Chapter 2: Science of Spice and Flavour

'Cilantro Love and Hate: Is It a Genetic Trait?', Shwu, 23andMe Research, 24 September 2012, https://blog.23andme.com/23andme-research/cilantro-love-hate-genetic-trait/ (last accessed on 25 June 2020).

'Q & A with Psychological Scientist Linda Bartoshuk', Association for Psychological Science, https://www.psychologicalscience.org/publications/observer/obsonline/q-a-with-taste-expert-linda-bartoshuk.html (last accessed on 25 June 2020).

'Reducing Sodium in Foods: The Effect on Flavor. Nutrients', D.G. Liem, Fatemeh Miremadi and Russell Keast (2011).

Essentials of Food Science, Vickie A. Vaclavik and Elizabeth W. Christian, Tad Campbell, Springer Nature (2014), p. 4.

'F Is for Flavor', Stella Culinary School, Jacob Burton, https://www.youtube.com/watch?v=Z9L-tJxPTGY (last accessed on 25 June 2020).

The Science of Spice: Understand Flavour Connections and Revolutionize Your Cooking, Stuart Farrimond, Dorling Kindersley Limited, London (2018), p. 13.

Chapter 4: Dropping Acid

'Where Is the Acid?', Science and Cooking Public Lecture Series 2014, Harvard University, https://www.youtube.com/watch?v=oqRRZD9OT0E (last accessed on 25 June 2020).

The Art of Flavour: Practices and Principles for Creating Delicious Food, Daniel Patterson and Mandy Aftel, Robinson (2018), p. 236.

Chapter 5: Umami, Soda, Rum

'Umami: Why the Fifth Taste Is So Important', Amy Fleming, *Guardian*, 9 April 2013, www.theguardian.com/lifeandstyle/wordofmouth/2013/apr/09/umami-fifth-taste (last accessed on 25 June 2020).

'The Science of Satisfaction', Sam Kean, Science History Institute, 8 October 2015, www.sciencehistory.org/distillations/magazine/the-science-of-satisfaction (last accessed on 25 June 2020).

'The Best Crispy Roast Potatoes Ever Recipe', J. Kenji López-Alt, Serious Eats, www.seriouseats.com/recipes/2016/12/the-best-roast-potatoes-ever-recipe.html (last accessed on 25 June 2020).

'Fish and Chips Recipe', Heston Blumenthal, *GQ* (Britain), 16 February 2016, www.gq-magazine.co.uk/article/fish-and-chips-recipe (last accessed on 25 June 2020).

Chapter 6: Taking It to the Next Level

'Sous Vide Time and Temperature Guide', ChefSteps.com, www.chefsteps.com/activities/sous-vide-time-and-temperature-guide (last accessed on 25 June 2020).

Chapter 7: Burn the Recipe

L'art De La Cuisine Française Au Dix-neuvième Siècle, Marie Antonin Carême, Adamant Media Corporation (2005).

The Everyday Healthy Vegetarian: Delicious Meals from the Indian Kitchen, Nandita Iyer, Hachette India (2018).

Chapter 8: The Biryani

'At 43 Orders Every Minute, Biryani Is the Most Sought-After Dish on Swiggy', Lata Jha, Livemint, 7 August 2019, www.livemint.com/news/india/at-43-orders-every-minute-biryani-is-the-most-sought-after-dish-on-swiggy-1565186078419.html (last accessed on 25 June 2020).

'Kolkata Mutton Biryani Recipe—Ramzan & Eid Special Recipe—Bengali-Style Mutton Biryani At Home', Bong Eats (YouTube), https://www.youtube.com/watch?v=SbWGXcZTYzg (last accessed on 25 June 2020).

Bibliography

1. *Modernist Cuisine: The Art and Science of Cooking*, Nathan Myhrvold, Chris Young and Maxime Bilet.

2. *Cooking for Geeks: Real Science, Great Hacks, and Good Food*, Jeff Potter.

3. *On Food and Cooking: The Science and Lore of the Kitchen*, Harold McGee.

4. *Keys to Good Cooking: A Guide to Making the Best of Foods and Recipes*, Harold McGee.

5. *The Science of Spice: Understand Flavour Connections and Revolutionize Your Cooking*, Stuart Farrimond.

6. *What Einstein Told His Cook: Kitchen Science Explained*, Robert L. Wolke.

7. *The Food Lab: Better Home Cooking Through Science*, J. Kenji López-Alt.

8. *The Kitchen As Laboratory: Reflections on the Science of Food and Cooking*, edited by Cesar Vega, Job Ubbink, Eric Van Der Linden.

9. *Salt, Fat, Acid, Heat: Mastering the Elements of Good Cooking*, Samin Nosrat.

10. *The Flavour Bible: The Essential Guide to Culinary Creativity*, Karen Page.

11. *Essentials of Food Science*, Vickie A. Vaclavik and Elizabeth W. Christian.

12. *Food: The Chemistry of Its Components*, Tom P. Coultate.

13. *Ratio: The Simple Rules behind the Everyday Craft of Cooking*, Michael Ruhlman.

14. *The Story of Our Food*, K.T. Achaya.

Internet Recommendations

YouTube Channels:

1. BongEats

2. Nisha Madhulika

3. Madras Samayal

4. Your Food Lab (by Sanjyot Keer)

Blogs:

1. Saffron Trail

2. Serious Eats

Twitter:

1. @KitchenChemProf

2. @ajit_bhaskar

3. @maxdavinci

Acknowledgements

I have to start by thanking my wonderful wife, Smitha, who has a fantastic palate that can detect way more aromas and tastes than mine, and who serves as the general arbiter for most of my kitchen experiments.

My mother, Rajeswari, for whom cooking was a stressful daily chore involving three hungry boys, all while managing a day job at a bank. She was my first mentor in the kitchen. My pictures of her cooking have their own cult following on Twitter (@krishashok).

My late father, G. A. Krishnan, who quietly encouraged me to expand my eating habits outside of the strictly no-garlic vegetarian confines of my childhood home.

My in-laws, Madhavan Nair, who is an aficionado of all my attempts at cooking Punjabi food thanks to his life in the army, and Parvathi Vijayalakshmi, whose fish curry the denizens of the Bay of Bengal will gladly give their lives to be a part of.

My younger brother, Krish Raghav, who curates things better than most people, for introducing me to some amazing food places around the world.

My network of like-minded cooking and food science enthusiasts on Twitter, without whose daily, illuminating conversations over the years this book would not have been possible. There are too many to name, but you know who you are.

My editor, Manasi, who pursued me for five years until I wrote this book.

My son, Samanyu, for every single hour that I spent writing this book during the Covid-19 lockdown, which I should have spent playing with him.

Index

A

Acetobacter aceti 146

acid/acids xvii xix–xx, xxi, xxii, xxv, 9–10, 12, 19, 20, 43– 44, 53, 65, 77, 81, 85, 87, 91, 92, 102, 119, 135–40, 152–53, 160–61, 162, 167, 175, 178, 182, 188, 191, 193, 195, 198, 201, 206, 212, 214, 217, 218, 219, 221, 229, 230, 231, 232, 235, 236, 237–38, 240–41, 249 (*see also under* separate entries) acetic 146; amino 65, 102, 107, 112, 115, 127, 128, 130, 156; citric 132, 138, 146; culinary xxv, 135, 146,149–151, 240; glutamic 65– 66, 156; hydrochloric xix, 12, 117, 136; lactic 44, 140, 153; liquid 153; malic 151; sulphuric xx, 112; tartaric 142, 151; unsaturated fatty 59–60, 61–62

Adams, Douglas 1

airtight containers 1, 87

ajinomoto. See monosodium glutamate (MSG)

alcohol xxi, xxi, 65, 81, 85, 132, 146, 149, 163–67, 229, 231; beer 137, 149, 163, 165, 167, 179; brandy xxi, 166, 232; rum 155–68; vodka xxi, 166, 167; wine 81, 139, 146–147, 164–65, 166–67

aldehydes 52, 6364, 65, 69, 72, 97, 102

alkaloids 65, 150

amchoor (powder form of dried green mangoes) xxii, 64, 87, 99, 100, 102, 116, 144–45, 152, 175, 191, 231, 236

amylopectin 11, 38–39, 202, 205, 207, 213, 246, 251, 257

anardana xxii, 149

appam 161

D

dal makhani xxii, 223, 232

deep-frying xxv, 15, 21, 25, 58, 60–61, 76, 122, 131–32, 263

defrosting 54, *see also* refrigerator

dehydrator 176

density 5, 44, 73

dish, dry 21, 51, 144, 188, 202

dosas 43–45, 116, 140, 161; batters for 161, *see also* fermentation

dough 22, 47–48, 55, 77, 131, 180, 193, 210–14 254, 262; nongluten-based 213

drumstick/moringa xiv

E

eggs xxiv, 21, 38, 52, 55, 107, 124, 131, 147, 161, 182, 193; boiled 55, 107, 147, 202; poached 203; science of 54–56; scrambled 203; yolks 181, 221, 240

electromagnetic: radiation 9, 258; waves 7, 169–70, 173

electronic pressure cookers xxvi, 142, 177–78; Instant Pot xxvi, 177

emulsifier 178, 180, 240

endorphins xxvii, 82–83

Eno 159

enzymes 9, 12, 50, 79, 95, 110, 112–113, 166

Espagnole 220

ethanol 146, 163–64, 167

F

G

K

N

O

P

tomatoes xv, xvi, xviii, xxi, xxii, 18, 19, 144, 147–49, 152, 156, 157, 158, 164, 166, 195–96, 218–20, 223, 238–39; acids and 147; paste xxi, 148, 196, 229; puree 148, 197; science of 147–49

tools xv, 22, 263

translucent xxi, 55, 106, 115, 118, 119, 173

turmeric xiv, 86, 87, 103, 172, 187, 222, 233

U

umami 65–68, 88, 125, 156–57, 158–59, 182, 244; ingredients 230

urad dal xvi–xvii, 43, 74, 118, 118,130, 160, 227, 263

utthapams 44–45

V

vadas 131–32, 161

vegetables xxiii, xxvii, 1, 14, 18, 43, 159, 160, 172, 195–98, 244–46, 249; English xviii; frozen 217; science of 48–51; season's raw 239

Velouté 220

vessels: aluminium 19; for cooking 19–23, 21, 31, 37, 41, 121, 147, 175, 254; material science of 19–23, 37; steel 5, 25, 141

vinegar xx, xxii, xxv, 119, 135, 138, 146–47, 148, 151, 153, 161, 229, 236, 238, 240; apple cider 146–47; red wine 146–47; science of 146–47

vinegar-pickled ginger xxii

vitamins 27, 38, 44, 107, 198